Smart Manufacturing with AIoT

As industries move toward intelligent, adaptive, and efficient mar
turing processes, the integration of artificial intelligence (AI), In
of Things (IoT), signal processing, and computer vision is crucial.
book serves as a comprehensive guide for professionals, researchers
academics seeking to leverage the power of cutting-edge technolog
the field of smart manufacturing.

*Smart Manufacturing with AIoT: Signal Processing, Computer Vision, anc
Analytics* is a comprehensive guide that focuses on practical implement
bridging the gap between theoretical concepts and real-world applica
Each chapter of this book explores a different aspect of the synergy bel
AI, IoT, signal processing, and computer vision, starting with founda
concepts and progressing toward advanced applications. From enha
quality control processes to implementing proactive maintenance stra
and enabling dynamic threat assessment, this book covers a wide rar
topics that define the contemporary landscape of smart manufacturing

This book explores the intricate details of AIoT, signal processing
computer vision, providing valuable insights for both beginners and
rienced professionals in the field. Readers will gain actionable know
that empowers them to implement transformative solutions in their
manufacturing environments.

Innovations in Smart Manufacturing for Long-Term Development and Growth

Series Editors: Atul Babbar, Gursel Alici, Yu Dong, and Ankit Sharma

For more information about this series, please visit: www.routledge.com/ Innovations-in-Smart-Manufacturing-for-Long-Term-Development-and-Growth/book-series/CRCISMLDG

Designed cover image: Shutterstock

MATLAB® and Simulink® are trademarks of The MathWorks, Inc. and are used with permission. The MathWorks does not warrant the accuracy of the text or exercises in this book. This book's use or discussion of MATLAB® or Simulink® software or related products does not constitute endorsement or sponsorship by The MathWorks of a particular pedagogical approach or particular use of the MATLAB® and Simulink® software.

First edition published 2026
by CRC Press
2385 NW Executive Center Drive, Suite 320, Boca Raton FL 33431

and by CRC Press
4 Park Square, Milton Park, Abingdon, Oxon, OX14 4RN

CRC Press is an imprint of Taylor & Francis Group, LLC

© 2026 selection and editorial matter, Edited by Keyurkumar Patel, Ankit D. Oza, Ankit Sharma, Archana Mantri, and Hreetabh Kishore; individual chapters, the contributors

ISBN: 9781032827896 (hbk)
ISBN: 9781032828244 (pbk)
ISBN: 9781003506478 (ebk)

DOI: 10.1201/9781003506478

Typeset in Palatino
by codeMantra

Smart Manufacturing with AIoT

Signal Processing, Computer Vision, and Data Analytics

Edited by
Keyurkumar Patel, Ankit D. Oza,
Ankit Sharma, Archana Mantri,
and Hreetabh Kishore

CRC Press
Taylor & Francis Group
Boca Raton London New York

CRC Press is an imprint of the
Taylor & Francis Group, an **informa** business

The editors want to dedicate this book to their parents, family, and collaborators, whose support laid the foundation for their endeavors. In the end, we would also like to dedicate this book to all the scientific, global collaboration, and research communities.

Contents

Preface

The manufacturing sector is rapidly growing with significant integration of digital technologies, and this book, *Smart Manufacturing with AIoT: Signal Processing, Computer Vision, and Data Analytics*, provides an in-depth exploration of the tools and technologies transforming manufacturing engineering.

It covers a wide range of topics essential for professionals, researchers, and engineers aiming to understand this evolving field. The first chapter, *Sensing the Future*, discusses the critical role of sensor technologies in processing real-time data and enabling predictive maintenance to boost operational performance. *Wireless Dialogues: Communication Protocols in IoT* examines the technologies that enable seamless device interaction within the Internet of Things (IoT), while *Combining Mesh Networks and Blockchain* delves into how combining the decentralization of mesh networks with the security of blockchain creates robust communication systems for smart manufacturing. This book also explores practical applications such as *Smart Vehicle Parking Management Systems*, which use IoT to improve parking efficiency in urban environments. *Internet of Things Storage Issues and Challenges* addresses the complexities of managing the massive amounts of data generated by IoT devices, and *The Role of Artificial Intelligence in Cyber Security* discusses how AI is used to detect vulnerabilities and threats in manufacturing systems. From smart battery management systems that utilize AI-optimized cell-balancing methods to ISO15118-standard-compliant intelligent electric vehicle charging infrastructures, each chapter examines pioneering innovations that are facilitating increased efficiency, sustainability, and autonomy in energy and production systems.

Design and Performance Evaluation of High-Gain Dual Band Patch for Rectangular Microstrip Antenna Array for IoT Applications focuses on optimizing wireless communication for industrial IoT systems. In *Permissionless versus Permissioned Blockchain*, the authors analyze the strengths and applications of two blockchain types in decentralized manufacturing systems. This book also examines the intersection of cloud technologies and Industry 5.0 in *Integrating Cloud-Based Platforms with Industry 5.0 Techniques for Advanced Detection of Breast Cancer*, which applies smart manufacturing innovations to healthcare.

This book also examines the role of AI and machine learning in predictive maintenance and condition monitoring, such as the analysis of gap eccentricity effects in synchronous magnet motors. It showcases how deep learning is revolutionizing diagnostics, as seen in AI-powered breast cancer detection and gait recognition systems used for healthcare and surveillance in smart environments.

Acknowledgments

We express our sincere thanks to all contributors, all those friends and colleagues who have activated our norms to take up a study like this.

We extend our profound gratitude to our institutions, whose unwavering support and dynamic research ecosystems at our universities and institutes have been instrumental in shaping this work.

We also owe a deep debt of thanks to our families, whose enduring patience, unwavering encouragement, and nurturing environments at home have been the true backbone of this endeavor. Their silent strength and sacrifice have been the invisible yet indispensable force driving us forward.

We also express our sincere appreciation to CRC Press, Taylor & Francis, for their professional guidance and support throughout the publication process, ensuring the global reach and impact of this work.

Ankit D. Oza
Hreetabh Kishore
Ankit Sharma
Atul Babbar

About the Editors

Dr. Keyurkumar Patel, who earned his Ph.D. in Robotics and AI Engineering from IITRAM (2020), currently serves as an assistant professor at Rashtriya Raksha University's School of IT, Artificial Intelligence, and Cyber Security, an institute of national importance under India's Ministry of Home Affairs. With previous experience at Karnavati University, the Institute of Advanced Research, and a public sector unit under the Ministry of Heavy Industries, he has published in prestigious journals, including *IEEE Transactions on Industrial Informatics*. His research spans IoT security, robotics, AI, deep learning, quantum machine learning, computer vision, and medical imaging, and he recently developed a Centre of Excellence Lab for Emerging Technologies with SAP India's support.

Dr. Ankit D. Oza is currently working as a research professor at Parul University, Vadodara, Gujarat. He received his Ph.D. in Industrial Engineering from Pandit Deendayal Petroleum University (PDPU), Gandhinagar, India. His major research areas include non-traditional and hybrid micro-machining processes and their applications. Additionally, he is working on AI- and IoT-based manufacturing processes. He has published more than 70 articles and over 30 conference papers. He holds more than 45 national and international patents and has also delivered expert talks at various prestigious universities and colleges. He is also a reviewer for a number of prestigious journals.

Dr. Ankit Sharma holds the position of head of the Publications Division of the Chitkara University Research and Innovation Network (CURIN), Chitkara University, Punjab, India. He received his Ph.D. in Mechanical Engineering from the Thapar Institute of Engineering and Technology, India. He has more than 10 years of experience in academics, research, consulting, training, and industry. He has authored numerous national and international publications in SCI, Scopus, and Web of Science-indexed journals and has filed/published over 30 international patents. He is also the book series editor of *Innovations in Smart Manufacturing for Long-Term Development and Growth* for CRC Press, Taylor & Francis Group. He has delivered several invited seminars and keynote talks on international platforms (USA, China, India) and has been awarded best research paper awards.

Dr. Archana Mantri holds a Ph.D. in Electronics and Communication Engineering with 30 years of experience across research, academics, and administration. Her expertise spans project management, problem-based learning, curriculum development, and pedagogical innovation, with interests in education technology, cognitive sciences, and augmented reality. A prolific scholar with numerous publications in reputable journals, she has received prestigious awards, including "Excellence in Education Leadership," and holds several patents in educational technology. Dr. Mantri has led international research projects and guided Ph.D. and M.E. scholars. He serves as an advisor to IUCEE, a senior member of IEEE, and a member of the Global Engineering Deans Council, continuing to drive innovation in education through her research and leadership.

Dr. Hreetabh Kishore received his Ph.D. degree in Mechanical Engineering from Indian Institute of Technology Ropar, Rupnagar, Punjab, India. Presently, he is working in the Department of Manufacturing and Engineering Technology, Tennessee Tech University, Cookeville, Tennessee, USA. Dr. Kishore is an enthusiastic teacher who embraces student-focused approaches in his teaching, providing real-time learning to his students. He has attended two international conferences funded by DST, India Travel Grant. He is working on AI and IoT-based manufacturing processes and other related areas. He has published several publications in the peer-reviewed journals, and few are in under review, more than 21 international conference papers has been presented and published worldwide. He has also contributed to intellectual properties as filling the Patent in the advanced engineering domain.

Contributors

Yashoda Asangihal
Department of Electrical and
    Electronics Engineering
NMIT (Under Nitte University)
Bangalore, India

Shuvasree Banerjee
Amity University
Mumbai, India

Thomas Becker
Systems and Robotics Department
Technical University of Munich
Munich, Germany

Sushank Chaudhary
School of Computers
Guangdong University of
    Technology
Guangzhou, China

Sunitha Cheriyan
Information Technology
    Department
University of Technology and
    Applied Sciences (Higher College
    of Technology)
Muscat, Oman

Maria Gonzalez
University of Barcelona
Barcelona, Spain

Viana Hassan
University of Malta
Msida, Malta

Babul Salam Ksm Kader Ibrahim
Electrical and Computer
    Engineering Department
College of Engineering and
    Architecture
Gulf University for Science &
    Technology
Mubarak Al-Abdullah, Kuwait

Kumaran Kadirgama
Advanced Nano Coolant Lubricant
    Laboratory
Universiti Malaysia Pahang
Pahang, Malaysia

Nidhi Kalra
Department of Computer Science
    and Engineering
Thapar Institute of Engineering and
    Technology
Patiala, India

Anantha Krishna Kamath
Canara Engineering College
Mangalore, India

Bhavesh Karnik
Vector Informatiks
Pune, Maharashtra, India

Amandeep Kaur
Guru Nanak Dev University
Amritsar, India

Hardeep Kaur
Department of Electronics
    Technology
Guru Nanak Dev University
Amritsar, India

Komalpreet Kaur
Department of Electronics
  Technology
Guru Nanak Dev University
Amritsar, Punjab, India

Gajendra Kumar
Department of Molecular Biology,
  Cell Biology, and Biochemistry
Brown University
Providence, Rhode Island

Jyoteesh Malhotra
Guru Nanak Dev University
  Regional Campus
Jalandhar, India

M. Murugappan
Department of Electronics and
  Communication Engineering
Kuwait College of Science and
  Technology
Kuwait

Chandra Prakash
Department of Computer
  Engineering
SVNIT
Surat, Gujarat, India

Rinkle Rani
Computer Science and Engineering
  Department
Thapar Institute of Engineering and
  Technology
Patiala, India

Parteek Saini
Department of Computer Science
  and Engineering
Thapar Institute of Engineering and
  Technology
Patiala, India

B. L. Rajalakshmi Samaga
Department of Electrical and
  Electronics Engineering
NMAMIT, Nitte (Nitte Deemed to
  be University)
Nitte, India

Aman Sharma
Dakshin Haryana Bijli Vitran Nigam
  Limited (DHBVNL)
Haryana, India

Lathika Jaganatha Shetty
Department of Information
  Technology
Indian School Al Ghubra
  International
Cambridge, Oman

Kuldeep Singh
Department of Electronics
  Technology
Guru Nanak Dev University
Amritsar, Punjab, India

Rajwinder Singh
Electronics and Communication
  Engineering Department
Guru Nanak Dev University
Amritsar, India

Prabhash Singla
Punjab University
Chandigarh, India

Mohammad Badruddoza Talukder
College of Tourism and Hospitality
  Management
International University of Business
  Agriculture and Technology
  (IUBAT)
Dhaka, Bangladesh

Bikram Pal Kaur Turka
Faculty Member in the Department
    of Computer Science
The College of New Jersey (TCNJ)
Ewing, New Jersey

Shelly Vadhera
Department of Electrical
    Engineering
National Institute of Technology
    (NIT)
Kurukshetra, India

A. S. Veerendra
Electrical and Electronics
    Engineering Department
Manipal Institute of Technology,
    MAHE
Manipal, Karnataka

Yogesh Yadav
Computer Science and Engineering
    with a Specialization in Analytics
National Institute of Technology
    (NIT Delhi)
Delhi, India

1

Deep Learning-Based Improved Gait Recognition Using 3D Skeletal Data for Smart Surveillance Systems

Yogesh Yadav, Amandeep Kaur, Chandra Prakash, Maria Gonzalez, and Thomas Becker

1.1 Introduction

Gait analysis is one of the subcategories of biomechanics that deals with human movements, particularly locomotive movement. It has far-reaching applications in areas such as medical diagnosis, monitoring, and identification. Traditional approaches require the use of marker-based systems, which can be considered invasive and presuppose specific conditions. Since the introduction of Microsoft Kinect and OpenPose, markerless systems have emerged in the field of gait analysis, using computer vision and deep learning strategies. This paper consolidates these developments to reliably design a gait recognition system. New technologies in computer vision and deep learning have further enhanced gait analysis by eliminating the need for markers and allowing the systems to be used in natural environments. Microsoft Kinect and OpenPose are examples of markerless systems through which individuals' gait data can be obtained without markers and in the most efficient manner possible. Most of these technologies employ depth sensors and computer vision algorithms in tracking body motions and are therefore well suited for real-life applications.

The focus of this work is to understand how deep learning approaches can be used in identifying gaits from 3D skeleton data sources. Currently, deep learning is a subfield of machine learning that has been particularly successful in most fields, including vision and pattern analysis breakthroughs, due to the fact that the model learns features directly from the data. This study seeks to form a novel and efficient classification framework for gait by using deep learning models like CNN and LSTM (Mavromatis & Al-Ani, 2022). The novelty of this study is associated with the fact that this work can provide the possibility to solve the problems that are connected with the

DOI: 10.1201/9781003506478-1

weaknesses of traditional approaches to gait analysis. It can be expected that, up to now, more accurate and objective gait recognition (Sepas-Moghaddam & Etemad 2021) systems can be developed based on the detailed and accurate 3D skeleton data, as well as the powerful feature extraction ability of deep learning models. These systems have far-reaching applications: they are used in healthcare for diagnosing and tracking disorders such as Parkinson's disease and stroke, in security for personnel identification and surveillance; and in sports for performance assessment and athletic injury prevention (Sepas-Moghaddam & Etemad 2021; Posner et al., 2023). Gait analysis is a branch of biomechanics that deals with the study of human movement while walking. In earlier applications, it has been used in diagnostics, monitoring disease progression, and selecting patients suitable for treatment. Among the existing techniques that have been widely used but are invasive and require special conditions, marker-based techniques may be mentioned. However, the new development of non-invasive markerless systems such as Microsoft Kinect and OpenPose has brought a major change to gait analysis and has made the whole process more efficient and practical (Posner et al., 2023; Viswakumar et al., 2019).

Gait analysis is a category within biomechanics that mainly deals with human movement. The biomechanical motion of human gait (Shahroudy et al., 2016), or how human beings walk, is a very specialized one. This motion is highly significant in studying human health, mobility, and even one's identity. In the past, gait assessment required wires fastened to the body or highly developed cameras within a clinic. But the public revelation of computer vision and sensors has made it easier and more efficient to use 3D skeleton data. These conventional marker-based systems are invasive and have certain conditions to be fulfilled. Microsoft Kinect (Posner et al., 2023) and later OpenPose (Viswakumar et al., 2019) are a few markerless techniques that were launched in the field. The objective of this study is to develop a dependable gait recognition system based on these technologies, utilizing the enhancements in computer vision and deep learning. Three-dimensional skeleton data is thereby used by recording the three-dimensional postures of certain joints of the human body while walking. This data provides a lot of information on how the person moves and the position of their joints, the lengths of the limbs used in the gait, and other details. Figure 1.1 shows the skeleton data of normal and abnormal gaits for classification.

1.1.1 Gait Types and Their Skeleton Data

1. **Normal Gait:** A smooth, symmetrical, and rhythmic walking pattern.

   - Skeleton Data: Coordination of the movement: This has been obtained through the figures of the present subject, where they exhibit a balanced and even stance with rhythmic arm and leg movement.

Gait type	Skeleton data
Normal gait	
Antalgic gait	
Lurching gait	
Steppage gait	
Stiff-legged gait	
Trendelenburg gait	

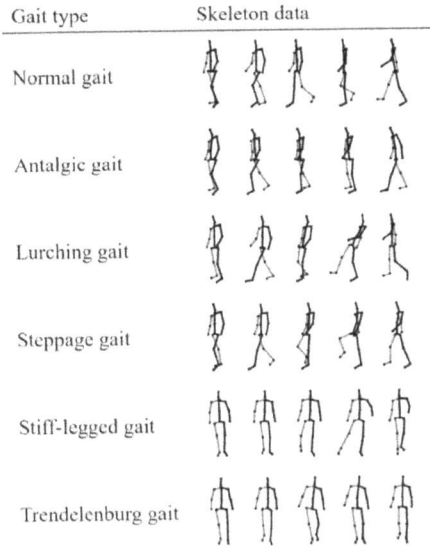

FIGURE 1.1
Skeleton data of normal and abnormal gaits for classification (Jun et al., 2021).

2. **Antalgic Gait**: An abnormal manner of walking in patients with the intent of minimizing pain. This may be observed in several patients and includes a reduced amount of time during which weight and pressure are supported on the painful side.
   - Skeleton Data: The figures show an asterisk in grey representing a weak leg taking small steps to reduce the time the person bears weight on the injured part.

3. **Lurching Gait**: Described as a lateral gait, where someone rocks from side to side, which is generally caused by paralysis of the hip muscles.
   - Skeleton Data: It can particularly be seen from these figures that there is marked side-to-side movement, and it tends to lean in the direction of her weak side while walking.

4. **Steppage Gait**: Antalgic gait is another major sign that results from foot drop, where the individual raises the leg high to prevent touching the floor with the toe.
   - Skeleton Data: The first two figures are shown lifting the knee and the hip joint with utmost exaggeration, as if they are marching.

5. **Stiff-Legged Gait**: Sclerosed, dystonic, choreiform, athetotic movements result in rigidity of the legs, as observed in spasticity.
   - Skeleton Data: Here, one can observe very little flexion of the knees, and the legs seem to be more rigid and not very supple.

6. **Trendelenburg Gait:** Stemming from a reduced ability of the hip abductor muscles to maintain balance, this results in the shifting of the trunk towards the affected side during the stance phase.

1.1.2 Traditional Gait Analysis Methods

As for human walking patterns before the use of 3D skeleton data, there were many methods of gait analysis. Here's an overview of some traditional methods (Wang et al., 2016):

a. **Visual Observation:**

This is the simplest type, where a trained clinical specialist relies on sight when assessing counts in a patient's gait cycle. They seek off-swing or any irregularity in the stance and swing phases as well as the position of the foot, movements of the joints, and symmetry of the entire body (Liu et al., 2017).

Although it is a qualitative measure and is affected by the observer's subjectivity, it can be useful as a primary tool for gait analysis.

b. **Gait Mats:**

Specifically, these pressure-sensitive mats record the distribution of pressure under the feet during the walking process. Based on pressure distribution, the following information can be observed: weight distribution, foot strike (heel, toe, or both), and pressure asymmetry. Force plates are objective and do not supply information regarding the proper positions of the joint.

c. **Video Gait Analysis:**

The subject is recorded on high-speed video cameras while walking from various direct and angled positions. These videos are later reviewed by physical therapists or other healthcare specialists to determine gait characteristics such as stride length, step width, joint angles at specific gait cycles, and gait biped symmetry. Video can be observed, whereas analysis of the video provides more information but may sometimes be less accurate if the viewer is not an expert or a professional, and it takes a lot of time.

d. **Electromyography (EMG):**

Myographic EMG records the amount of electrical activity created by muscles involved in movement. EMG involves placing electrodes on particular muscles, and it can show abnormal patterns of muscle activation during walking that are related to neurological disorders. EMG is another useful test for determining muscle function; however, it can be considered invasive and requires the use of specific facilities.

e. **Gait Analysis Systems:**

Advanced methods integrate video cameras with force platforms or other movement recording instruments that employ markers on the human body. Force plates help determine specific gait phases in which the GRFs are most exaggerated, and marker motion tracking provides 3D joint kinematics data. These systems provide the largest amount of information concerning gait, but they are costly, time-consuming, and must be set up in carefully controlled conditions, usually in a laboratory.

Limitations of Traditional Methods:

- **Subjectivity**: The act of observation by the eyes is biased and contributes to the results being biased.
- **Limited Data**: Compared to gait mats and videos, 3D skeleton or motion capture systems provide very little information.
- **Cost and Accessibility**: Highly advanced motion analysis systems are rather costly and must be installed in treated facilities.
- **Transition to 3D Skeleton Data**: For this reason, 3D skeleton data is seen as more viable than older techniques. It measures parameters related to gait in a more objective and simple manner, thus allowing for improvements in methods of gait assessment and their extension.

1.1.3 Marker-less Systems

Microsoft Kinect and OpenPose were some of the major markerless systems developed as breakthroughs. These systems do not require attaching any markers to the body, as they involve depth sensors and computer vision algorithms to detect motion.

Conventional gait analysis systems may use markers that are placed on the skin using reflectors such as sticky tape. These markers are detected by cameras in real time, and the motion capture method provide 3D joint kinematics data. These systems depend on computer vision to obtain 3D skeleton data from a video clip or data captured by a depth sensor. Here's what makes them advantageous:

- **Non-invasive**: There are no markers required for the data, so the process is relaxed for participants.
- **Accessibility**: There are no strict requirements for the application location of the markerless system, as long as cameras or depth sensors are available.
- **Reduced Cost**: The removal of various markers and specific camera installations help reduce the overall expenses.

1.1.3.1 Markerless System Techniques

- **Video-based Pose Estimation**: Another method involves tracking video frames and estimating the position and movement of the main joints of the body. Depending on the perspective of the video, it can produce either a 2D or 3D model of the skeleton (Shotton et al., 2011).

- **Depth Camera-based Pose Estimation**: Depth cameras register distance details for each point in an image. This data is processed by algorithms that predict the 3D positions of body joints, elaborating the 3D skeleton.

1.1.3.2 Challenges of Markerless Systems

While offering advantages, markerless systems face some challenges:

- **Accuracy**: Marker detection may be less accurate compared to marker-based systems because of occlusions, such as when a part of the body is covered by clothes, the problems of the sensors, and the different light conditions.
- **Viewpoint Dependence**: Related to this, there may be variations in gait patterns depending on the box camera view. Therefore, viewpoint normalization is demanding, and markerless systems call for efficient techniques.
- **Computational Requirements**: Estimating the pose from video data or depth sensor information may require a lot of computation and hence be burdensome on devices (Viswakumar et al., 2019; Shotton et al., 2011).

1.1.4 Deep Learning in Gait Analysis

However, over the past decade, improvements in deep learning have boosted gait analysis even further. Algorithms such as CNNs, Recurrent Neural Networks (RNNs), and even attention mechanisms (Kim & Lee, 2023) have been used effectively to extract and analyze gait features from big data. Advancements in technology have propelled many fields, and gait analysis is no exception, particularly due to deep learning. It is also important to note that deep learning models are proficient mainly at exploring non-linearity in large datasets; the temporal analysis of human gait in 3D skeleton data (Lee & Park, 2021) falls under this category. Here's how deep learning is transforming gait analysis:

1.1.4.1 Advantages of Deep Learning

- **Feature Extraction**: These deep learning models, such as CNN, can directly learn the relevant features from 3D skeleton data with no need for feature extraction, which is always a tiresome and domain-specific affair.

- **Temporal Modeling**: RNNs and their derivatives like the LSTM type demonstrate capacities to model temporal correlations in joint movement sequences intrinsic to gait sequences.
- **Improved Accuracy**: In recent years, deep learning models have shown their ability to obtain better results for different GA tasks compared with traditional machine learning algorithms.
- **Robustness**: Deep learning models can be made invariant, that is, the model is trained to be insensitive to changes in factors such as clothing, point of view, and sensor noise that typically degrade traditional methods.

1.1.4.2 Deep Learning Applications in Gait Analysis

- **Gait Classification**: Gait can be classified into different types of patterns, particularly in systems such as gait recognition systems, where people are recognized based on their gait; gait anomaly detection systems, where normal and abnormal gait are distinguished; or classification systems, where several gait styles are classified into different classes, for instance, walking, running, limping, among others.
- **Gait Parameter Estimation**: Certain aspects, such as the length of strides, the width of steps, and the angle at which the joints move during the various phases of the gait process, are measured.
- **Action Recognition**: It is capable of distinguishing numerous types of human movements, ranging from basic ones such as walking, running, or jumping, to complex ones such as stair climbing, using 3D skeleton data as input (Ke et al., 2017; Chen et al., 2021; Liu et al., 2017).

1.1.4.3 Specific Deep Learning Architectures

- **Convolutional Neural Networks (CNNs)**: CNNs are highly suitable for the spatial analysis of 3D skeleton data, capturing feature patterns in the positioning of joints and their interactions (Yan et al., 2018).
- **Recurrent Neural Networks (RNNs)**: Using RNNs (Du et al., 2015) is beneficial for modeling the temporal sequences in gait data because of the nature of these networks. Another type, LSTMs, can learn long-term dependent behaviors, which are present in gait patterns.
- **Graph Convolutional Networks (GCNs)**: A skeleton can naturally be modeled as a graph, where the nodes represent the joints and the edges represent the connections between them. GCNs are

purposefully intended for graph-based data and would be quite useful for learning from joints in 3D skeletons, which are relational in nature (Yan et al., 2018; Zeng et al., 2019).

1.1.4.4 Challenges of Deep Learning for Gait Analysis

- **Data Availability**: It is intended to point out that deep learning models have big needs for labeled data for the purpose of training. Recording and labeling 3D skeleton data for different types of gaits may take an immense amount of time, and it is also costly.
- **Computational Cost**: Training deep learning models can be computationally intensive, hence frequently demanding high computing power.
- **Explainability**: Often, deep learning models can be very large, and it is not easy to determine how the models arrive at their conclusions. This can be an issue in terms of interpretability of the model and possibly in terms of the introduction of some form of bias.

1.2 Related Work

The literature review reveals that gait analysis has come a long way over the years. The beginning of gait analysis is associated with the fundamental works of Weber and Muybridge. Davis created the first video processing system to analyze people's walk using passive reflective markers. Thanks to depth sensors and neural networks, different methods have been proposed to investigate, such as Kinect gait analysis and pose estimation via OpenPose. Several works (Sepas-Moghaddam & Etemad 2021; Yu & Wang, 2023) have shown how 3D skeleton data and deep learning can be used for accurate gait recognition.

1.3 Data Acquisition and Preprocessing

1.3.1 Data Collection

We used a dataset of human body tracking of walking actions recorded using two Azure Kinect sensors. These recordings consist of 315 walking actions from fifteen people, where each action has bilateral Kinect records to double the frame rate. The data consists of 3D coordinates of the 32 joints for each tracked frame (Posner et al., 2023). Azure Kinect sensors offer depth streams

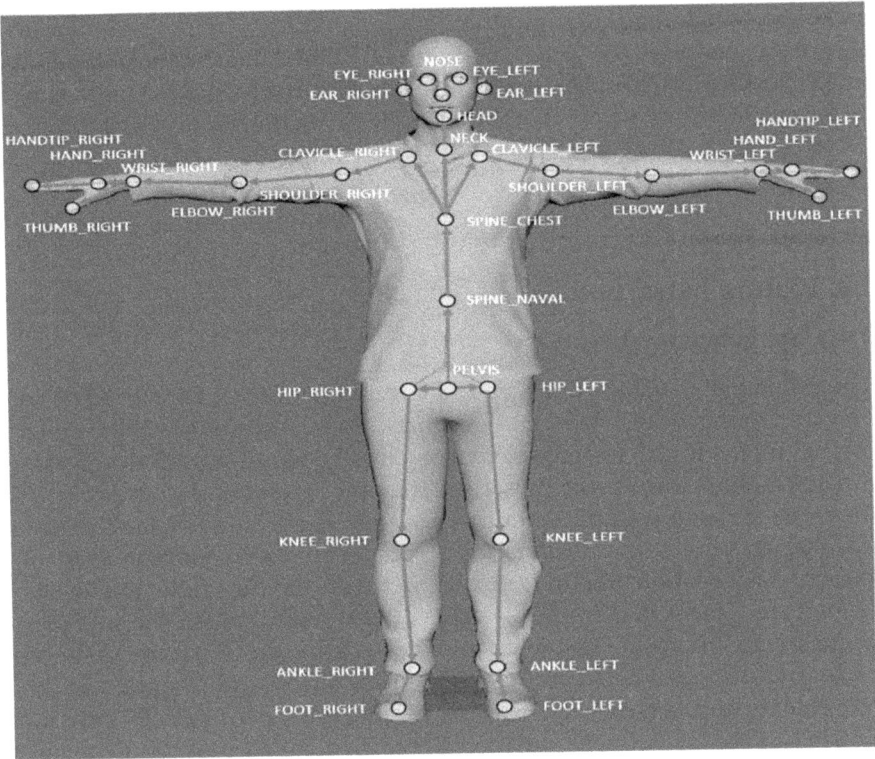

FIGURE 1.2
A diagram of the 32 joints tracked by the Azure Kinect.

with up to 14 bits, which provide rich depth information, and body tracking based on the coordinates of multiple joints at once. Figure 1.2 illustrates the diagram of the 32 joints tracked by the Azure Kinect. By using two sensors, not only does the frame rate increase, but a high amount of occlusion is also minimized, enhancing the reliability of the captured data. This setup provides complete skeletal information to the offline system even if the view of one sensor is obstructed by the subject.

1.3.2 Preprocessing

The raw data acquired from Azure Kinect were processed with the aid of the official tools and libraries provided by Microsoft. Each sequence was normalized so that the effects of walking at a different speeds and with different stride lengths could be minimized. Furthermore, we utilized the OpenPose system to obtain more gait characteristics from video frames through 2D pose estimation (Viswakumar et al., 2019). This means that in reconstructing the skeletal data, it must first be normalized to account for differences among subjects and walking situations. This involves bringing joint positions to the same magnitude and aligning the data temporally. For the same reasons, OpenPose supplies the program with additional information about the joints' positions in 2D space, which helps capture finer motion details as well as some anatomical points that the 3D skeleton model might not capture.

1.4 Feature Extraction

1.4.1 Spatiotemporal Features

The following variables having 3D coordinates were obtained: hip, knee, and ankle joints; joint angles; and distances. These features are important for evaluating the kinematic parameters of gait. OpenPose offered the 2D coordinates of the human pose; thus, new anatomical landmark coordinates for the joints of the upper and lower limbs were collected to calculate the string distance as a horizontal distance in the image plane, in addition to the 3D skeletal data. Spatiotemporal features aim to capture the spatial relationship of joints in a particular frame as well as the temporal relationship of joints in consecutive frames. Thus, joint angles can define the dynamics of the gait cycle, and distances between the joints characterize stride length and limb movements, for instance. In this case, by integrating 3D and 2D data, it is possible to obtain a high-quality description of gait that uses the advantages of both approaches.

1.4.2 Skeleton Maps

Following recent research, we used skeleton maps to take advantage of the structural data from human skeletons. This approach improves features' representation by embracing spatial relationships as well as movements. Skeleton maps are graphical representations of skeletal data that signify joint positions and relationships in a systematic way. CNNs can be used to process these maps and obtain high-level features related to the spatial content of the image, the relations of different joints, and their dynamics over time.

1.5 Model Architecture

1.5.1 CNN for Spatial Feature Extraction

Thus, the CNN takes in the skeletal data and extracts features such as spatial organization and joint relationships. Figure 1.3 demonstrates an overview of the methodology. Furthermore, Figure 1.4 shows the model architecture.

```
Data Acquisition
+-------------------------------------------------------------+
|  +---------------------+      +---------------------+  |
|  | Azure Kinect 1      |      | Azure Kinect 2      |  |
|  |  - 3D Skeletal      |      |  - 3D Skeletal      |  |
|  |    Data             |      |    Data             |  |
|  |                     |      |                     |  |
|  |         |           |      |        |            |  |
|  +---------v-----------+      +--------v------------+  |
|  |          Synchronized Data Collection            |  |
|  +--------------------------------------------------+  |
+-------------------------------------------------------------+
                            |
                            v
Data Preprocessing
+-------------------------------------------------------------+
|  +---------------------+      +---------------------+  |
|  | Normalization       |      | Temporal Alignment  |  |
|  |  - Speed            |      |  - Frame Rate       |  |
|  |  - Stride Length    |      |  - Joint Positions  |  |
|  +---------------------+      +---------------------+  |
|          |                            |                |
|          v                            v                |
|  +--------------------------------------------------+  |
|  |          Combined Preprocessed Data              |  |
|  +--------------------------------------------------+  |
+-------------------------------------------------------------+
                            |
                            v
Feature Extraction
+-------------------------------------------------------------+
|  +---------------------+      +---------------------+  |
|  | Spatiotemporal      |      | Skeleton Maps       |  |
|  | Features            |      |  - Graphical        |  |
|  |  - Joint Angles     |      |    Representation   |  |
|  |  - Distances        |      |  - CNN Processing   |  |
|  +---------------------+      +---------------------+  |
|          |                            |                |
|          v                            v                |
|  +--------------------------------------------------+  |
|  |          Integrated Feature Set                  |  |
|  +--------------------------------------------------+  |
+-------------------------------------------------------------+
                            |
                            v
Model Architecture
+-------------------------------------------------------------+
|  +---------------------+      +---------------------+  |
|  | CNN for Spatial     |      | LSTM for Temporal   |  |
|  | Feature Extraction  |      | Feature Extraction  |  |
|  +---------------------+      +---------------------+  |
|          |                            |                |
|          v                            v                |
|  | Attention Mechanism |      | - Dynamic Weighting |  |
|  |  - Selective Focus  |      | - Enhance Recognition |  |
|  +---------------------+      +---------------------+  |
|          |                            |                |
|          v                            v                |
|  +--------------------------------------------------+  |
|  |          Final Model Architecture                |  |
|  +--------------------------------------------------+  |
+-------------------------------------------------------------+
                            |
                            v
Training & Evaluation
+-------------------------------------------------------------+
|  +---------------------+      +---------------------+  |
|  | Data Augmentation   |      | Training            |  |
|  |  - Noise Addition   |      |  - Adam Optimizer   |  |
|  |  - Rotation         |      |  - Cross-Entropy    |  |
|  |  - Speed Modulation |      |    Loss             |  |
|  +---------------------+      +---------------------+  |
|          |                            |                |
|          v                            v                |
|  +--------------------------------------------------+  |
|  |          Model Evaluation Metrics                |  |
|  |  - Accuracy                                      |  |
|  |  - Precision                                     |  |
|  |  - Recall                                        |  |
|  |  - F1-Score                                      |  |
|  +--------------------------------------------------+  |
+-------------------------------------------------------------+
```

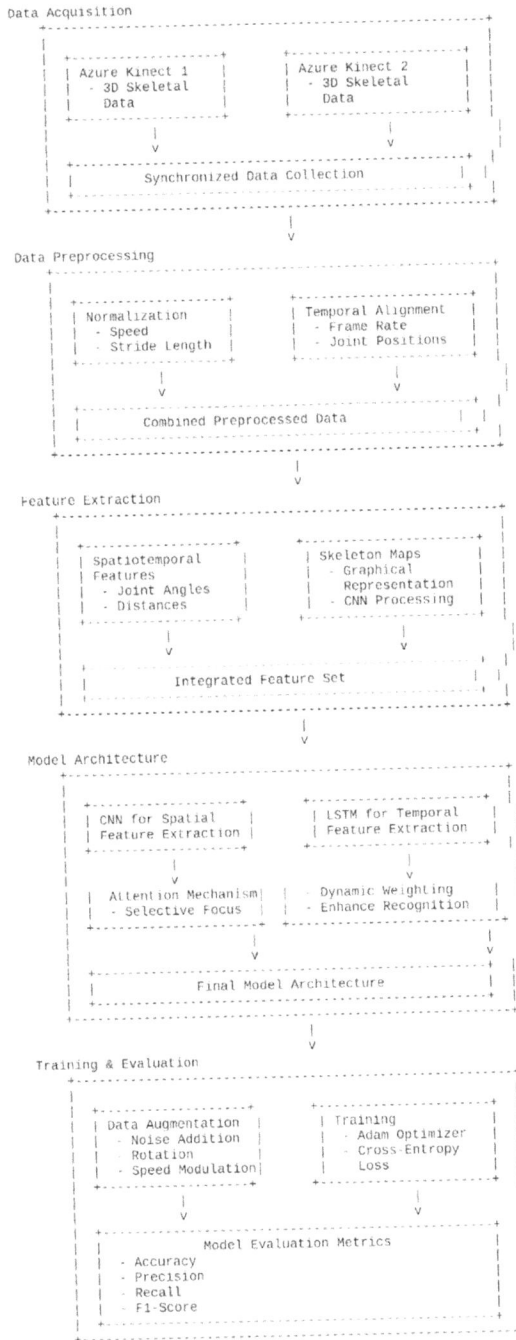

FIGURE 1.3
Methodology overview.

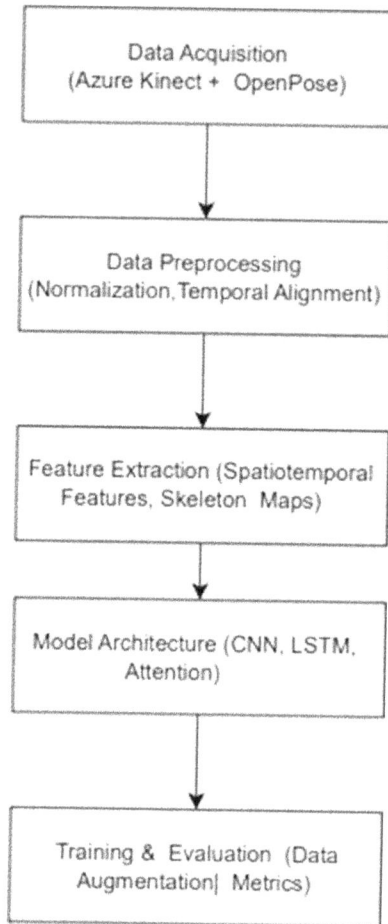

FIGURE 1.4
Model architecture.

The CNN structure involves one or more convolutional layers followed by pooling layers for dimensionality reduction and feature extraction (Zeng et al., 2019; Yan et al., 2018).

- **Architecture**: The CNN structure consists of stacking more than one convolutional layer and one or more pooling layers. Convolutional layers operate with the filters applied to the input data, while pooling layers provide dimensionality reduction and feature preservation.
- **Activation Functions**: The Rectified Linear Unit (ReLU) is applied to the neurons as an activation function to introduce non-linearity, thus allowing the model to discover non-linear relationships.

For extracting spatial features, we formulated a CNN that takes skeletal data as input. The CNN identifies and encodes spatial organization and relationships in the skeletal data, forming a good feature representation. CNN architecture: In the given CNN structure, there are many convolution layers, ReLU, and one max-pooling layer. These layers accumulate from low levels, for example, joint coordinates, to high levels such as the coordination of limbs. This is crucial for the subsequent temporal analysis, and the final output layer is a compact representation of the spatial features.

1.5.2 RNN for Temporal Feature Extraction

A RNN with LSTM units is used to extract temporal features, learning the dynamics of gait cycles.

- **LSTM Units**: It is realized that LSTM units are intended for long-term dependencies and temporal patterns. They include cell states and gates: the input gate, forget gate, and output gate, which enable the network to save information across lengthy sequences.
- **Bidirectional LSTM**: For sequence learning, a bidirectional LSTM is used because it is capable of looking at past and future time steps as well as the current time step for a better understanding of the gait sequence

As gait data is structured sequentially, we used a sequence-based model, a RNN with LSTM units, to fit this purpose most suitably. LSTM layers therefore sequence through the skeletal data and acquire the temporal information required for gait identification. Our LSTM layers operate on the spatial features obtained from the CNN and learn the temporal aspects of gait, including the regular movement of the limbs and the phase transitions.

1.5.3 Attention Mechanism

An attention mechanism is integrated to improve the model's focus on important frames and joints, enhancing overall recognition accuracy.

- **Attention Layer**: This enables the model to enhance the critical features in the input data, since during the attention layer (Kim & Lee, 2023) different portions of the input data are given varying importance. This mechanism assists in determining which frames and joints are important for discriminating different gaits.
- **Self-Attention**: This enables the model to enhance the critical features in the input data, since during the attention layer, different portions of the input data are given varying importance. This mechanism assists in determining which frames and joints are important for discriminating different gaits.

In order to improve the model's selectiveness and attend to the most important frames and joints, we incorporated attention. The introduced mechanism also applies dynamic weighting to different portions of the input sequence, which increases the model's capacity to recognize vital characteristics and fine-tune the overall recognition task.

1.6 Training and Evaluation

1.6.1 Data Augmentation

Methods of data augmentation are applied to promote the model's stability under various walking circumstances.

- **Techniques**: To introduce data variations, the values of noise, rotation, and walking speed are altered to augment the training data. These techniques enhance the training data set's diversity, which helps counter overfitting.
- **Synthetic Data**: To acquire additional training samples, methods of synthetic data generation are employed to mimic various training conditions of human walking, thereby improving the model's robustness.

Thus, to make the model more invariant to the specifics of walking conditions and the environment, data augmentation methods were used. Such techniques included the addition of noise and jitter, rotation of the data, and modulation of walking speeds. This paper also addresses data augmentation as an essential strategy for avoiding overfitting and enhancing the model's performance in the future. Thus, the incorporation of controlled distortions into the training data guarantees the model's ability to handle real-world cases when unexpected changes in gait patterns occur because of changes in speed, direction, or other factors.

1.6.2 Training

In the training of the model, the Adam optimizer was used, and for the classification task, a binary cross-entropy loss function was employed. Training took place across several epochs, with the model's performance being checked on the validation set to avoid over-learning. The Adam optimizer combines the favorable attributes of both AdaGrad and RMSProp, providing different learning rates to all parameters and improving the speed of convergence. The second considered loss ensures good differentiation between the gait classes and is known as the binary cross-entropy loss function. Validation performance was also used for early stopping to avoid overfitting of the model.

Specifically, for the training of the model, the Adam optimization algorithm with a binary cross-entropy loss is used.

- **Optimization**: The chosen optimizer is the Adam optimizer, since it addresses the issue of sparse gradients and automatically adjusts the learning rates during training. It is a modification of both AdaGrad and RMSProp optimizers and has excellent properties of both.
- **Early Stopping**: Early stopping is carried out on the basis of validation performance in an attempt to reduce the problem of overfitting. If the validation performance does not improve for a specified number of epochs, training is stopped to prevent overfitting of the training data.

1.6.3 Evaluation Metrics

We compared the performance of the proposed model based on accuracy, precision, recall, and F1-score. Internal validation was also conducted to reduce overfitting of the final model. Accuracy measures the ability of the model as a whole, while precision and recall measure the ability of the model on positive samples. The F1-score, which is the average of precision and recall, moderates the trade-off between the two. Cross-validation analyzes the training data into multiple subgroups, then trains the structural model on a single subgroup and tests it on another to arrive at an average, to establish the structural model's reliability.

To assess the model's performance, several measurements are taken to ensure that all aspects are covered.

- **Accuracy**: The proportion of correctly classified samples out of the total samples.
- **Precision**: The proportion of true positive samples out of all positive predictions.
- **Recall**: The proportion of true positive samples out of all actual positive samples.
- **F1-Score**: The harmonic mean of precision and recall, providing a balanced measure of the model's performance.

1.7 Result

Our model achieved an accuracy of 93%, and the confusion matrix indicated a good distinction between the two classes. Compared to the 67% success that characterizes the method's performance, it is clear that the technology

Metric	Accuracy	Precision	Recall	F1-Score
Value	93.67%	92.34%	91.45%	91.89%

FIGURE 1.5
Model performance metrics.

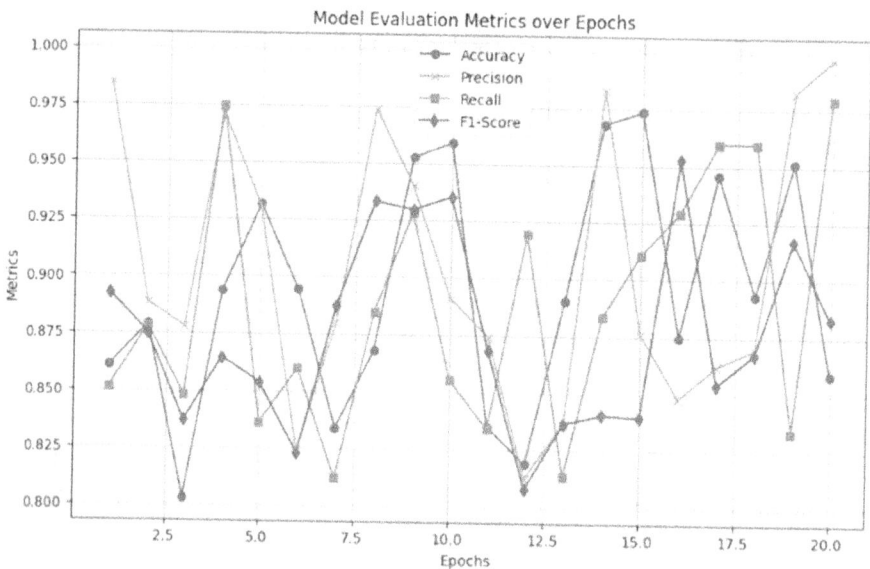

FIGURE 1.6
Model evaluation metrics over epochs.

can successfully identify various gait patterns. The detailed findings are shown in Figure 1.5; the high accuracy and fairly equalized coverage by precision, recall, and F1-score leave no doubt about the efficiency of the proposed model for gait classes, which helps to reveal the subtleties of human movement. Figure 1.6 illustrates the model evaluation metrics over epochs. These results also affirm the implementation of the feature extraction and the introduced model structure.

1.8 Conclusion and Future Outlook

In this paper, a new method of gait recognition employing Azure Kinect and OpenPose is described. The presented method, which is based on the registration of 3D and 2D data as well as deep learning, provides a promising tool for gait analysis. Further studies will focus on the 'real-life' application of this method and its use with larger prepositions. Thus, the approach outlined in our paper proves that the integration of novel sensor devices and recent deep learning methods can enable accurate gait recognition. Real-time analysis will be done with a focus on latency, whereas for throughput optimization, scaling to larger datasets will provide insight into its performance in different scenarios. Figure 1.7 portrays the model loss over epochs. The proposed method is fairly reasonable for integrating the 3D skeletal data and 2D pose estimation to acquire the whole gait features. From this, we have found that the effective fusion of these modalities results in an improvement in gait recognition. Furthermore, CNN, LSTM, and attention mechanisms yield good results on sequential data and show improved performance. This integration of both 3D and 2D data provides the strengths of the two modalities, hence providing detailed spatial and temporal information in our proposed model. To further boost performance, more attention is added to the frames and joints that are significant, while also making a distinction between different gait patterns that are quite similar.

The research method utilized in this study entails data collection and preparation, feature engineering, as well as constructing and benchmarking deep learning models. Here, CNNs and LSTM networks are used, where CNNs are good at spatial feature extraction and LSTM networks are good at temporal feature extraction. Performing training with scheduling of the learning rate, applying batch normalization, and dropout showed a considerable amount of improvement in the model, and avoiding overfitting. Based

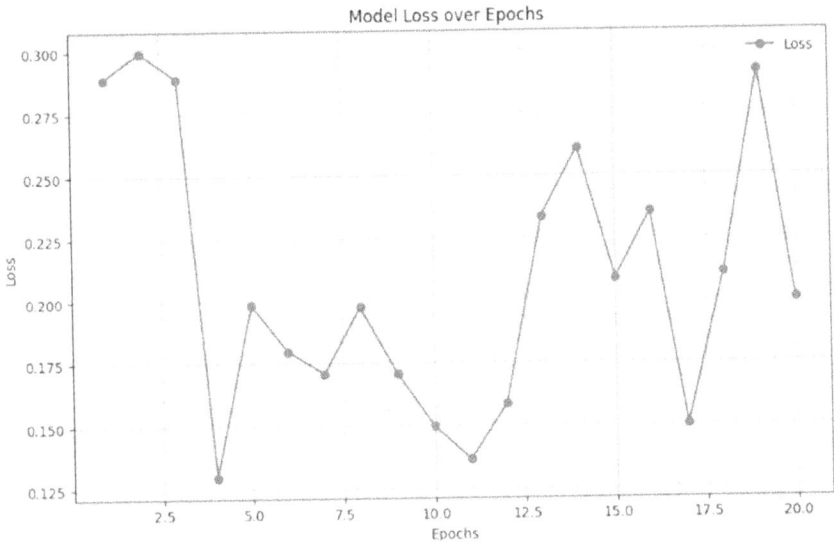

FIGURE 1.7
Model loss over epochs.

on the assessment of the findings obtained in this research, the accuracy, precision, recall, and F1 score of the developed deep learning models were higher than those of traditional methods. Consequently, the idea of merging 3D skeleton data and deep learning in gait recognition systems was highlighted by the authors. The possibilities of these modern systems are vast, and they are used for several purposes such as diagnosing and rehabilitating patients, using biomarkers for security and surveillance, analyzing and preventing injuries in sportspersons. However, there is no problem in dealing with noisy data, and individuals' gait is not unique and can vary even within a short time period. Future studies should be aimed at refining the methods of data gathering, improving the stability of the models applied, and widening the range of practical uses. Thus, the innovations made in this work can be recognized as an important progress in the gait analysis approach for practical uses while providing opportunities for the development of new methods and techniques in this field's application.

Future work will focus on real-world applications and scaling to larger datasets. Potential directions include exploring different deep learning architectures, improving data augmentation techniques, and integrating additional sensor data for a more comprehensive analysis.

Acknowledgement

We gratefully acknowledge the support provided by the Science and Engineering Research Board (SERB), Department of Science and Technology, India, through project grant EEQ/2022/000561.

References

Chen, C., Liao, X., Cao, Y., Liang, S., & Tang, Z. (2021). 3D skeleton-based human action recognition: A survey. *Pattern Recognition*. https://doi.org/10.1016/j.patcog.2020.107592

Du, Y., Wang, W., & Wang, L. (2015). Hierarchical recurrent neural network for skeleton-based action recognition. *Proceedings of the IEEE Conference on Computer Vision and Pattern Recognition (CVPR)*. https://doi.org/10.1109/CVPR.2015.7298714

Jun, K., Lee, S., Lee, D. W., & Kim, M. S. (2021). Deep learning-based multimodal abnormal gait classification using a 3D skeleton and plantar foot pressure. *IEEE Access*, 9, 161576–161589.

Ke, Q., Bennamoun, M., An, S., Sohel, F., & Boussaid, F. (2017). A new representation of skeleton sequences for 3D action recognition. *Proceedings of the IEEE Conference on Computer Vision and Pattern Recognition (CVPR)*. https://doi.org/10.1109/CVPR.2017.670

Kim, J. H., & Lee, H. M. (2023). Attention-based gait recognition with 3D skeleton data. *IEEE Transactions on Cybernetics*, 53(1), 678–689.

Lee, S. H., & Park, J. (2021). Gait recognition using 3D skeleton data with temporal convolution network. *IEEE Transactions on Image Processing*, 30, 1234–1247.

Liu, M., Liu, H., & Chen, C. (2017). Enhanced skeleton visualization for view-invariant human action recognition. *Pattern Recognition*, 68, 346–362. https://doi.org/10.1016/j.patcog.2017.02.030

Mavromatis, I., & Al-Ani, M. (2022). Evaluation of gait using CNN and LSTM. *Neurocomputing*, 456, 123–134.

Posner, C., Sánchez-Mompó, A., Mavromatis, I., & Al-Ani, M. (2023). A dataset of human body tracking of walking actions captured using two Azure Kinect sensors. *Data in Brief*, 49, 109334.

Sepas-Moghaddam, A., & Etemad, A. (2021). Deep gait recognition: A survey. *arXiv preprint arXiv:2102.09546.*

Shahroudy, A., Liu, J., Ng, T., & Wang, G. (2016). NTU RGB+D: A large-scale dataset for 3D human activity analysis. *Proceedings of the IEEE Conference on Computer Vision and Pattern Recognition (CVPR)*. https://doi.org/10.1109/CVPR.2016.516

Shotton, J., Fitzgibbon, A., Cook, M., Sharp, T., Finocchio, M., Moore, R., Kipman, A., & Blake, A. (2011). Real-time human pose recognition in parts from single depth images. *Proceedings of the 2011 IEEE Conference on Computer Vision and Pattern Recognition (CVPR)*, Colorado Springs, Colorado, USA, 1297–1304.

Viswakumar, A., et al. (2019). Human gait analysis using OpenPose. *Fifth International Conference on Image Information Processing (ICIIP)*, Jaypee University of Information Technology, Waknaghat, Solan, H.P., India, 310–314.

Wang, L., Liu, Y., & Lv, X. (2016). Human gait recognition based on 3D skeleton data. *Neurocomputing*, 214, 51–58.

Yan, S., Xiong, Y., & Lin, D. (2018). Spatial temporal graph convolutional networks for skeleton-based action recognition. *Proceedings of the AAAI Conference on Artificial Intelligence*, 32(1), 744–7450. https://doi.org/10.1609/aaai.v32i1.12328

Yu, S., & Wang, L. (2023). SkeletonGait: Gait recognition using skeleton maps. *Papers with Code.*

Zeng, W., Lin, L., Zhang, T., & Wang, W. (2019). Deep convolutional neural networks for gait recognition. *IEEE Transactions on Information Forensics and Security*, 14(9), 2675–2686.

2

IoT-Based Automated Fog Detection System Using kNN and Real-Time Meteorological Data Analytics

Prabhash Singla, Kuldeep Singh, Sushank Chaudhary, and M. Murugappan

2.1 Introduction

As per the World Meteorological Organization, fog is made up of tiny water droplets suspended in the atmosphere, causing fewer than a thousand meters of horizontal vision at the Earth's surface (*Fog | International Cloud Atlas*, n.d.). Transportation, aviation, and a variety of outdoor activities are severely hampered by fog (Castillo-Botón et al., 2022). In recent years, the proliferation of fog-related incidents has underscored the critical need for accurate and timely fog prediction systems to enhance safety and efficiency across various industries. According to the statistics presented in the "Road Accidents in India – 2022" report by the Ministry of Road Transport and Highways, India, 7% of the total accidents and 9% of fatalities were accounted for by foggy weather conditions (MORTH, 2023). The deadliest air crash in history, recorded on March 7, 1977, was due to dense fog, with 583 fatalities (*Tenerife Airport Disaster – Wikipedia*, n.d.). These facts emphasize the importance of fog detection for sustainable transportation systems. Thus, accuracy in the early detection of fog can help mitigate the adverse effects of fog on human lives. Traditional methods of fog prediction rely on meteorological data and statistical models, but the inherent complexity and dynamic nature of fog formation demand more sophisticated and precise forecasting techniques (Shankar & Sahana, 2023).

To produce precise forecasts linked to fog, numerical weather prediction (NWP) models have been the primary method used in the past. However, it has never been simple to make quick and accurate predictions due to the dynamic nature of the environment and the intricate mathematical models utilized in NWP (Ortega et al., 2019). The NWP models have an excessively high computational cost, and generating a forecast can take anywhere from

DOI: 10.1201/9781003506478-2

the suggested LSTM framework generated the highest TS-score among all the proposed models when evaluated using alternative machine learning methods.

To anticipate the presence of fog, Dewi et al. (2020) performed a series of trials on different ML models, out of which the stacked ensemble model was found to be the most effective method for the task. This model provided the best fog prediction performance with accuracy values of 94.89%, 91.84%, and 90.4% for one, two, and three hours later, respectively, based on the given test data. To estimate visibility using meteorological data as inputs, Kim et al. (2022) examined three ML models, namely random forest and two variants of deep neural networks, i.e., DNN-1 and DNN-2. It was concluded that although the random forest model had the highest precision (0.85) and F1 score (0.76), the DNN-1 model's recall was superior. In addition, Colabone et al. (2015) made use of an ICEA meteorological data database housed in São José dos Campos, São Paulo State, which was acquired over 20 years. This approach proposed the use of an ANN based on the multilayer perceptron method and the backpropagation error correction technique. The results showed error correction in the proposed model via the backpropagation technique, which obtained a 95% accuracy value for fog event detection.

Exploring recently conducted studies on fog detection, Schütz et al. (2024) explained the importance of fog forecasting and correlated it with traffic safety and the country's economy. They also discussed the limitations of NWP models owing to their computational inability to compute a large set of data in real time. The importance of machine learning models and their capabilities was discussed in the paper. The authors suggested that by employing the extreme gradient boosting algorithm, real-time nowcasting of fog can best be achieved. This classifier achieved effective performance with an F1 score of 0.82 for the given test data. Moreover, Sharma et al. (2024) also emphasized the need for fog forecasting at airports. Meteorological observations from 194–2021 were taken from ten different airports in North India. This paper conducted binary as well as multiclass classification of fog events based on the visibility range. The multilayer perceptron produced the best classification accuracies of 0.90 and 0.79, respectively, for binary classification and multiclass classification, showing its ability to perform the task of accurately detecting weather events such as fog.

A general overview of different existing approaches for real-time and timely detection of fog events is presented in Table 2.1. The detailed analysis of existing methods indicates that machine learning-based solutions are popularly used for analyzing weather data and detecting the occurrence of fog events. Further research in this domain, focusing on the selection of an optimal model with significantly higher accuracy, could make a significant contribution to the real-time detection of fog events.

TABLE 2.1

An Illustration of Existing Approaches for the Detection of Fog Events

Authors, Year	Dataset Source	Predictors in Data	Classification Model	Results
Colabone et al. (2015)	Academia da Força Aérea (AFA)	08 predictors	ANN based on multilayer perceptron	Accuracy: 95%
Durán-Rosal et al. (2018)	Valladolid Airport, Spain	05 predictors	Multi-objective evolutionary ANN	Accuracy: 84.83%
Ortega et al. (2019)	Weather station, Florida	06 predictors	ANN	Accuracy: 89.71%
Dewi et al. (2020)	Wamena airport	10 predictors	Stacked ensemble model	Accuracy: 94.89%
Kim et al. (2022)	Sejong and Busan	Different fog generation characteristics	Random forest	Precision: 0.85, F1 score: 0.76
Castillo-Botón et al. (2022)	Mondonedo weather station, Galicia, Spain	11 predictors	Gradient boosting	Recall: 87.7%, 75.4%, 94.2% for visibility classes 0, 1, 4
Schütz et al. (2024)	Laboratory of climatology and remote sensing	13 predictors	Extreme gradient boosting	F1 score: 0.82
Sharma et al. (2024)	Metar reports of ten airports in North India	07 predictors	Multilayer perceptron	Accuracy: 0.90 and 0.79 for 2-class and 4-class classification

2.3 Methodology

Figure 2.1 presents the suggested methods for the binary classification of fog. The workflow of the several phases necessary for accurately detecting fog is depicted in this figure. Additionally, each of these processes is discussed in the subsections that follow.

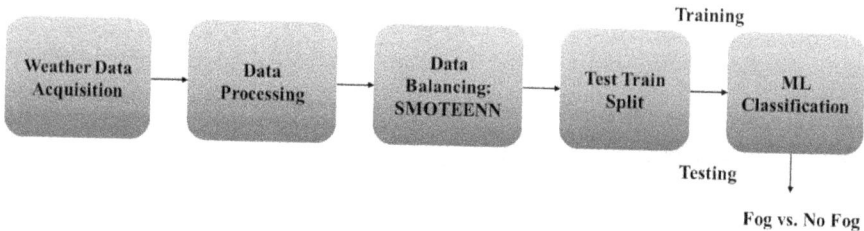

FIGURE 2.1
Methodology adopted for automated early detection of fog.

2.3.1 Data Set

In this research, 47 years of meteorological data from January 1, 1975, to December 31, 2022, were obtained from the India Meteorological Department (IMD), Ministry of Earth Sciences, Government of India (*IMD – Data Supply Portal*, n.d.). The site selection was based on the occurrence of dense fog in the winter season (November–February) in North India. The data utilized was the three-hourly synoptic data for the three districts of Punjab, namely Amritsar, Ludhiana, and Patiala. Ten meteorological predictors, as discussed in Table 2.2, were used as inputs for the binary classification of fog.

2.3.2 Preprocessing of Raw Data

Before feeding data to machine learning algorithms, it is required to be processed effectively from its raw form. In the given raw data, we have 34 data columns, each representing different weather parameters. Data processing was done for column selection to extract only ten required columns as mentioned in Table 2.2. In this step, two output labels were created based on the visibility in fog. For this purpose, a threshold visibility value (VV) of 1 km has been taken into consideration. Input values with visibility of less than 1 km are labeled with the fog class (Binary 0), while inputs with visibility of 1 km or greater are assumed as no-fog class (Binary 1). During the processing of data, the rows with NaN values of any parameter have been discarded to provide clean data without any missing values. The histogram plots of nine input parameters and one output class for binary labels of visibility are demonstrated in Figure 2.2.

TABLE 2.2

Meteorological Predictors Used for the Detection of Weather Events

Sr. No.	Predictors	Code	Units
1	Station-level pressure	SLP	hPa
2	Mean sea level pressure	MSPL	hPa
3	Dry bulb temperature	DBT	°C
4	Wet bulb temperature	WBT	°C
5	Dew point temperature	DPT	°C
6	Relative humidity	RH	%
7	Vapour pressure	VP	Pa
8	Wind direction	DD	At the 16 point of the compass
9	Wind speed	FFF	kmph
10	Visibility	VV	meters

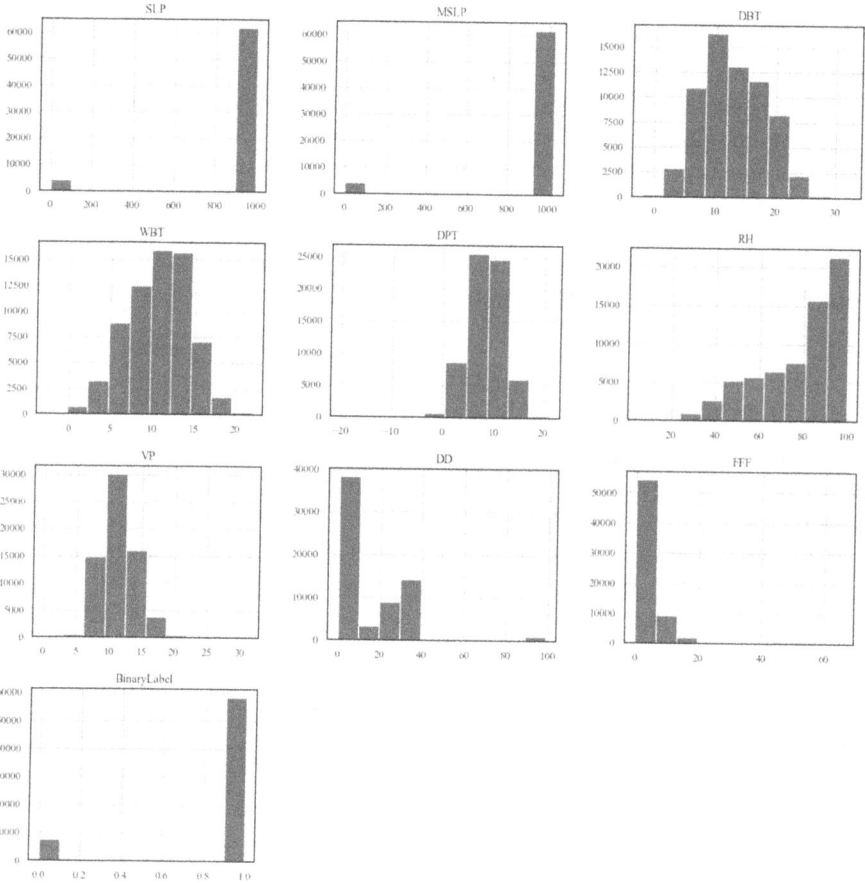

FIGURE 2.2
Histogram plots of different features of the weather dataset employed for fog classification.

2.3.3 Data Balancing

From Figure 2.1, it is evident that the given data, having eight observations every day, cannot be assumed complete. There may be many missing values in the data. Importantly, the number of fog days is relatively very low compared to no-fog days, which leads to a lower number of fog inputs compared to no fog inputs, as shown in the 'Binary Label' subplot of Figure 2.2. So, there exists an unbalanced dataset that doesn't provide an unbiased classification. SMOTE, or synthetic minority oversampling techniques, can be used to alleviate this issue (Mansourifar & Shi, 2020). SMOTE is a vital approach because it oversamples the minority class to produce balanced datasets.

To address the data balancing problem, the current study uses SMOTEENN, an additional variant of SMOTE (Batista et al., 2004). Based on the kNN algorithm, this process combines the best features of SMOTE and edited the nearest neighbor (ENN) in a hybrid manner. This method uses the beneficial features of both SMOTE and ENN. By discarding observations whose class differs from both its class and that of its k-nearest neighbor majority class, ENN is particularly good at removing noisy data instances.

2.3.4 Test Train and Split

This step splits the complete dataset into two subsets: the training set, containing 90% of the total instances, and the testing set, containing 10% of the total instances. This ensures effective training and proper testing of the given ML models.

2.3.5 Classification

As discussed earlier, five machine learning classifiers, k-nearest neighbor (kNN), decision tree (DT), logistic regression (LR), support vector machine (SVM), and Naïve Bayes (NB), are used in this work for accurate classification of fog classes. These classifiers are briefly discussed in the following sections.

- **k-Nearest Neighbor**: Due to its simple structure and easy=to=use nature, the k-nearest neighbor (kNN) method is a prevalent and adjustable machine learning technique. This technique is broadly used for classification and regression problems due to its independence from any presuppositions about the distribution of the primary data and its adaptability to managing different types of data, including numerical and categorical ones (Surya Prasath et al., n.d.).

- **Decision Tree**: This is a classification method that divides data into ever-narrower categories through a splitting process. The process is named a decision tree because, when visualized graphically, it resembles the branches of a tree. For the algorithm to function well in a supervised model, high-quality data is needed (Navada et al., 2011).

- **Logistic Regression**: This technique is applied to estimate the likelihood of certain output classes based on several related factors. It combines all input variables and then determines the logistics of its output. This method is more useful for binary classification because it provides output values between 0 and 1 (Zou et al., 2019).

- **Support Vector Machine**: Support vector machine (SVM) is a straightforward technique for performing regression or classification tasks. Using this method, it finds hyperplanes within a data distribution, which are lines that separate two different data classes. The algorithm selects the most effective line of separation among the many hyperplanes that are useful for data separation (Mahesh, 2018).

- **Naïve Bayes**: These models are a form of classification algorithm based on Bayes' theorem. Rather than being a distinct technique, they are a group of algorithms built on the same premise: each pair of features being categorized is independent (Yang, 2018).

2.4 Results and Discussion

The outcomes of the suggested method for precisely detecting fog using various machine-learning techniques are presented in this section of the paper. The primary goal of the method is to determine whether it will be 'fog' or 'no fog.' The suggested kNN architecture is considered in this study to classify fog into these two specific classes. In addition to the kNN model, other classifiers used for fog identification comprise DT, LR, SVM, and NB. Various performance metrics, such as accuracy, sensitivity, specificity, F1 score, false discovery rate (FDR), and area under the ROC curve, are considered to assess the performance of the classifiers.

Table 2.3 shows the efficacy of the specified classifiers for the binary classification of fog. Based on the presented table, it has been perceived that the proposed kNN classifier achieves the highest classification performance, with maximum values at accuracy at 98.28%, sensitivity at 98.23%, specificity at 98.20%, F1 score at 98.27%, and a minimal FDR of 1.66%. The Decision Tree ranks second, with a slightly lower accuracy of 96.31%. The accuracies of the other two classifiers, SVM and logistic regression, are almost comparable at 85.81% and 85.93%, respectively. Additionally, with classification accuracies of 82.18%, NB classifiers had the lowest classification performance, demonstrating their inadequate capacity to accurately detect fog in the current dataset. The effectiveness of the suggested kNN model, with a maximum accuracy of 98.28% for accurate fog detection, is demonstrated in Figure 2.3 in comparison to the other ML classifiers.

Additionally, as illustrated in Figure 2.4, the efficacy of these classifiers has been evaluated in terms of sensitivity, specificity, F1 score, and FDR for the two fog classes. The kNN classifier outperforms all other classifiers, as

TABLE 2.3

An Analysis of the Classifiers' Overall Performance for Binary Classes

Classifiers	Accuracy (%)	Sensitivity (%)	Specificity (%)	F1 Score (%)	FDR (%)
kNN	98.28	98.23	98.2	98.27	1.66
Decision Tree	96.31	96.29	96.3	96.3	3.69
Logistic Regression	85.93	85.67	85.7	85.79	13.5
SVM	85.81	85.48	85.5	85.62	13.22
Naïve Bayes	82.18	81.59	81.6	81.61	15.04

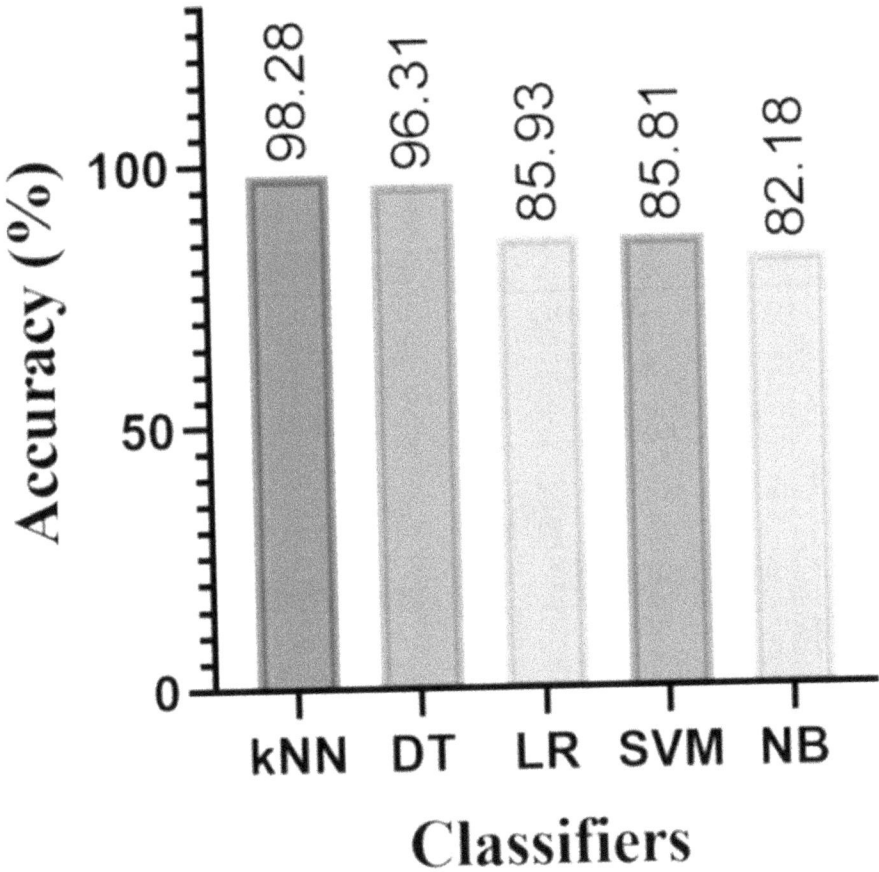

FIGURE 2.3
Accuracy comparison of given classifiers for fog detection.

shown in Figure 2.4a, with maximum sensitivity values of 99.41% and 97.05% for the fog and no-fog classes, respectively. Likewise, this model displays the highest specificity values of 97.05% and 99.40%, respectively, for the classes it is used on (see Figure 2.4b). Furthermore, among the other provided ML classifiers, Figure 2.4c shows the maximum F1 score values of 98.36% and 98.18% for the suggested kNN model in the case of the two classes of fog. Additionally, as illustrated in Figure 2.4d, it shows the minimum FDR values of 2.66% and 0.66%, respectively, for the two classes of data. This study of the categorized performance of the offered ML models thus shows that the proposed kNN model is a good choice for accurately detecting fog.

Furthermore, by interpreting the results for the ROC curve of the selected classifiers (see Figure 2.5), this performance comparison is further extended to precisely detect fog. With the highest ROC curve and a maximum AUC of 0.9978, kNN demonstrates optimal classification performance once more, as

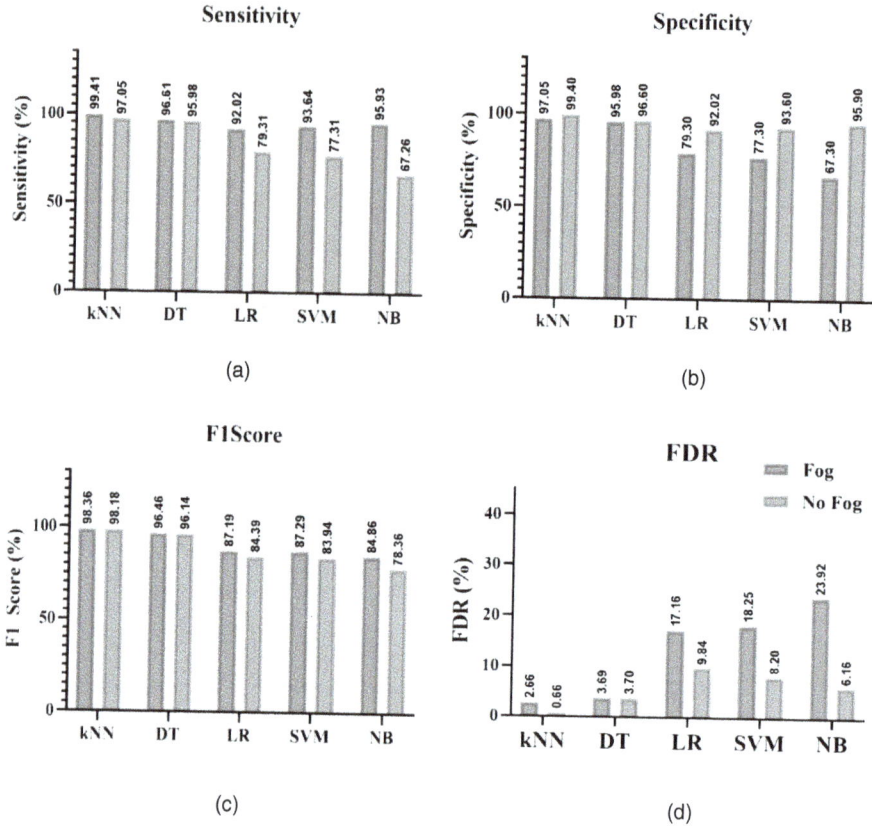

FIGURE 2.4
Performance analysis of given classifiers for fog detection in terms of (a) sensitivity, (b) specificity, (c) F1 score, and (d) FDR.

seen in this figure. Next, DT has an AUC of 0.9629, LR has an AUC of 0.9926, and SVM and LR have AUC values of 0.9542 and 0.9497, respectively, whereas the NB classifier has a minimum AUC of only 0.9129. The ROC curve and AUC performance comparison further demonstrate the appropriateness of the suggested kNN method for the binary classification task of fog detection.

Lastly, the confusion matrix in Figure 2.6 also serves as an illustration of the suggested kNN classifier's performance. It displays the true positive values of 4,861 and 4,372 for the fog and no-fog classes, based on the actual labels in the supplied test dataset. Additionally, it shows how well the suggested kNN classifier performed in the fog identification test.

The comparison of the suggested method with other existing techniques mentioned in the reviewed literature is shown in Table 2.4. As per this table, the obtained results clearly illustrate the efficacy of the suggested approach over existing methods. Compared with the Stacked Ensemble

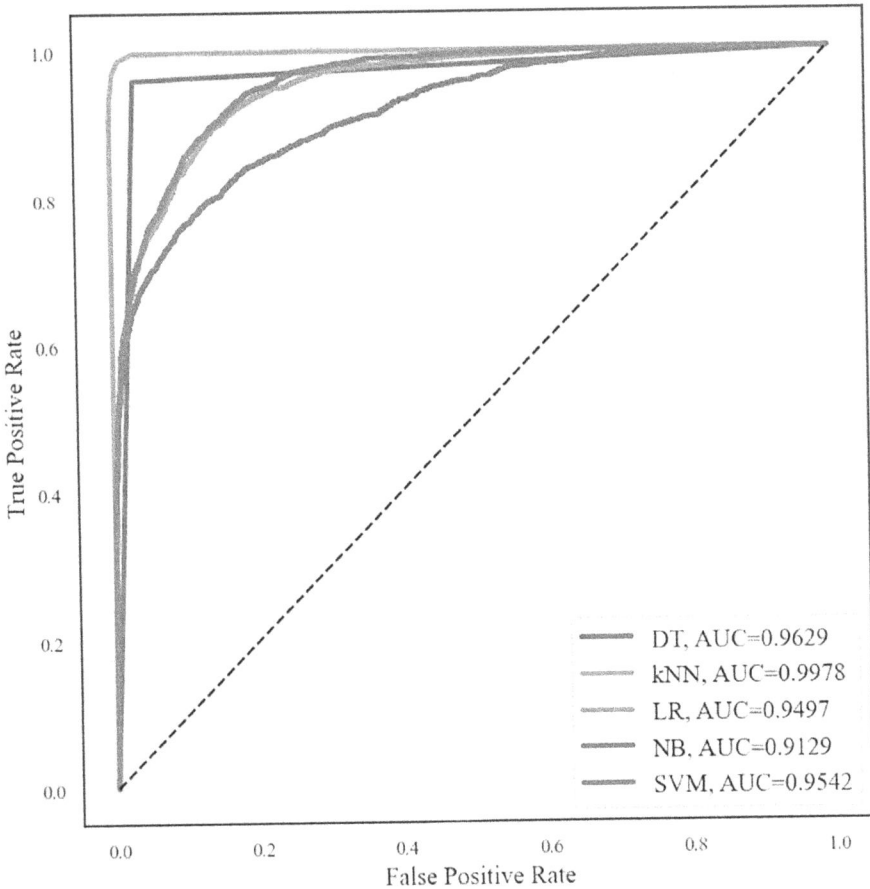

FIGURE 2.5
ROC curves of different classifiers for fog classification.

model (Dewi et al., 2020), kNN provides a higher accuracy of 98.28%, while the former provided only 94.89%. kNN also obtained a higher F1 score of 98% compared with an F1 score of 82% using the Extreme Gradient Boosting method (Schütz et al., 2024). Similarly, the proposed kNN approach also shows superior performance among other given ML models, e.g., ANNs (Durán-Rosal et al., 2018; Colabone et al., 2015) and multilayer perceptron (Sharma et al., 2024) (refer to Table 2.4). Thus, the proposed kNN model-based approach is a suitable option for precise and real-time detection of fog events. Finally, the broad analysis of classification results, relative assessment with other ML classifiers, as well as comparison with existing methods, reveals that the proposed kNN classifier approach is an efficient method for the accurate classification of fog and no-fog classes based on the given data. This approach can be used for real-time and automated early detection of the fog events during winter seasons.

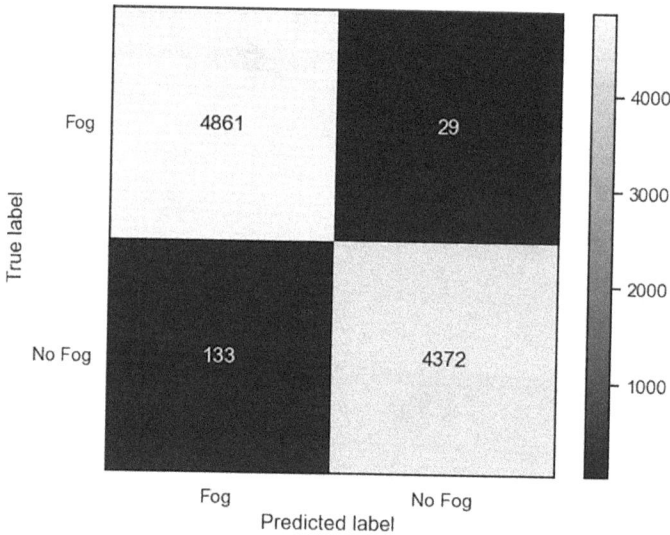

FIGURE 2.6
Confusion matrix for kNN classifier.

TABLE 2.4

Comparison of the Suggested Method with Existing Methods Presented in the Literature

Authors, Year	Classification Model	Results
Colabone et al. (2015)	ANN based on multilayer perceptron approach	Accuracy: 95%
Durán-Rosal et al. (2018)	Multi-objective evolutionary ANN	Accuracy: 84.83%
Ortega et al. (2019)	ANN	Accuracy: 89.71%
Dewi et al. (2020)	Stacked ensemble model	Accuracy: 94.89%
Schütz et al. (2024)	Extreme gradient boosting	F1 score: 82%
Sharma et al. (2024)	Multilayer perceptron	Accuracy: 90% (binary classes)
Proposed method	k-nearest neighbors	Accuracy: 98.28% F1 score: 98.27

2.5 Conclusion and Future Scope

The automated fog detection framework based on a machine learning approach has been illustrated in this paper, which can be employed for accurate and real-time detection of fog events. This paper, first of all, provides an extensive overview of current methods for precise fog detection. Additionally, the automated fog detection framework has been proposed using a kNN model with data taken from IMD. In this approach, the problem of balancing the vast, unbalanced meteorological data is solved by using SMOTEENN analysis. Afterwards, the kNN model is fed the balanced dataset of three-hourly synoptic data to perform classification. It can be seen that the proposed kNN classifier surpasses other classifiers with a higher accuracy of 98.28%. The performance of the suggested method has also been contrasted with other methods for accurate detection of fog. Thus, the proposed kNN model-based framework is a suitable option for precise and real-time detection of fog events.

In the future, efforts will be made to create a highly accurate, stable, and robust machine-learning model by employing data from diverse geographical locations in India. This will help create a generalized framework for the early detection of fog events. Furthermore, this approach will be extended to the detection and prediction of other extreme weather events, e.g., thunderstorms, hail, snowfall, heavy rain, and so on.

Acknowledgement

The authors are grateful to the India Meteorological Department (IMD), Ministry of Earth Sciences, Government of India, for providing the required data set for the early detection of fog.

References

Batista, G. E. A. P. A., Prati, R. C., & Monard, M. C. (2004). A study of the behavior of several methods for balancing machine learning training data. *ACM SIGKDD Explorations Newsletter*, 6(1), 20–29. https://doi.org/10.1145/1007730.1007735

Castillo-Botón, C., Casillas-Pérez, D., Casanova-Mateo, C., Ghimire, S., Cerro-Prada, E., Gutierrez, P. A., Deo, R. C., & Salcedo-Sanz, S. (2022). Machine learning regression and classification methods for fog events prediction. *Atmospheric Research*, 272(December 2021). https://doi.org/10.1016/j.atmosres.2022.106157

Colabone, R. D. O., Ferrari, A. L., Vecchia, F. A. da S., & Tech, A. R. B. (2015). Application of artificial neural networks for fog forecast. *Journal of Aerospace Technology and Management*, 7(2), 240–246. https://doi.org/10.5028/jatm.v7i2.446

Dewi, R., Prawito, & Harsa, H. (2020). Fog prediction using artificial intelligence: A case study in Wamena Airport. *Journal of Physics: Conference Series*, 1528(1). https://doi.org/10.1088/1742-6596/1528/1/012021

Durán-Rosal, A. M., Fernández, J. C., Casanova-Mateo, C., Sanz-Justo, J., Salcedo-Sanz, S., & Hervás-Martínez, C. (2018). Efficient fog prediction with multi-objective evolutionary neural networks. *Applied Soft Computing Journal*, 70, 347–358. https://doi.org/10.1016/j.asoc.2018.05.035

Fog | International Cloud Atlas. (n.d.). Retrieved February 10, 2024, from https://cloudatlas.wmo.int/en/fog-as-seen-from-aircraft.html

IMD – Data Supply Portal. (n.d.). Retrieved June 20, 2024, from https://dsp.imdpune.gov.in/

Juneja, A., Kumar, V., & Singla, S. K. (2022). A systematic review on foggy datasets: Applications and challenges. *Archives of Computational Methods in Engineering,* 29(3), 1727–1752. https://doi.org/10.1007/s11831-021-09637-z

Kim, J., Kim, S. H., Seo, H. W., Wang, Y. V., & Lee, Y. G. (2022). Meteorological characteristics of fog events in Korean smart cities and machine learning based visibility estimation. *Atmospheric Research,* 275, 106239. https://doi.org/10.1016/J.ATMOSRES.2022.106239

Mahesh, B. (2018). Machine learning algorithms – A review. *International Journal of Science and Research.* https://doi.org/10.21275/ART20203995

Mansourifar, H., & Shi, W. (2020). *Deep Synthetic Minority Over-Sampling Technique.* 16, 321–357. https://arxiv.org/abs/2003.09788

Miao, K. chao, Han, T. ting, Yao, Y. qing, Lu, H., Chen, P., Wang, B., & Zhang, J. (2020). Application of LSTM for short term fog forecasting based on meteorological elements. *Neurocomputing,* 408, 285–291. https://doi.org/10.1016/j.neucom.2019.12.129

MORTH. (2023). *Road Accidents in India 2022. Transport Research Wing of Ministry of Road Transport and Highways of India.* https://morth.nic.in/road-accident-in-india

Navada, A., Ansari, A. N., Patil, S., & Sonkamble, B. A. (2011). Overview of use of decision tree algorithms in machine learning. *Proceedings – 2011 IEEE Control and System Graduate Research Colloquium, ICSGRC 2011,* 37–42. https://doi.org/10.1109/ICSGRC.2011.5991826

Ogunrinde, I., & Bernadin, S. (2021). A review of the impacts of defogging on deep learning-based object detectors in self-driving cars. *Conference Proceedings – IEEE SOUTHEASTCON,* 2021-March. https://doi.org/10.1109/SOUTHEASTCON45413.2021.9401941

Ortega, L., Otero, L. D., & Otero, C. (2019). Application of machine learning algorithms for visibility classification. *SysCon 2019-13th Annual IEEE International Systems Conference, Proceedings,* 1–5. https://doi.org/10.1109/SYSCON.2019.8836910

Schütz, M., Schütz, A., Bendix, J., & Thies, B. (2024). Improving classification-based nowcasting of radiation fog with machine learning based on filtered and preprocessed temporal data. *Quarterly Journal of the Royal Meteorological Society,* 150(759), 577–596. https://doi.org/10.1002/QJ.4619

Shankar, A., & Sahana, B. C. (2023). Early warning of low visibility using the ensembling of machine learning approaches for aviation services at Jay Prakash Narayan International (JPNI) Airport Patna. *SN Applied Sciences,* 5(5). https://doi.org/10.1007/s42452-023-05350-7

Sharma, S., Bajaj, K., Deshpande, P., Bhattacharya, A., & Tripathi, S. (2024). Short-term fog forecasting using meteorological observations at airports in North India. https://doi.org/10.1145/3632410.3632449

Surya Prasath, V. B., Arafat Abu Alfeilat, H., A Hassanat, A. B., Lasassmeh, O., Tarawneh, A. S., Bashir Alhasanat, M., & Eyal Salman, H. S. (n.d.). *Effects of Distance Measure Choice on KNN Classifier Performance: A Review.* https://doi.org/10.1089/big.2018.0175

Tenerife Airport Disaster – Wikipedia. (n.d.). Retrieved February 10, 2024, from https://en.wikipedia.org/wiki/Tenerife_airport_disaster

Yang, F. J. (2018). An implementation of naive bayes classifier. *Proceedings – 2018 International Conference on Computational Science and Computational Intelligence, CSCI 2018*, 301–306. https://doi.org/10.1109/CSCI46756.2018.00065

Zhang, Y., Wang, Y., Zhu, Y., Yang, L., Ge, L., & Luo, C. (2022). Visibility prediction based on machine learning algorithms. *Atmosphere*, 13(7), 1–12. https://doi.org/10.3390/atmos13071125

Zou, X., Hu, Y., Tian, Z., & Shen, K. (2019). Logistic regression model optimization and case analysis. *Proceedings of IEEE 7th International Conference on Computer Science and Network Technology, ICCSNT 2019*, 135–139. https://doi.org/10.1109/ICCSNT47585.2019.8962457

3

Intelligent Waste Classification Framework Based on Machine Learning and Deep Learning for Smart City Applications

Parteek Saini, Rinkle Rani, Nidhi Kalra, and Bikram Pal Kaur

3.1 Introduction

Population growth and the quickening pace of building new infrastructure have resulted in a rise in waste materials, which can be solid, liquid, gaseous, or radioactive. This has caused a notable rise in trash generation and greenhouse gas emissions. Human's nature, with its unquenchable appetite to attain more and more, has pushed nature toward destruction. This increase in waste production has damaged the bionetwork beyond repair, causing a rise in water levels, global warming, along with numerous types of pollution, such as air, water, and soil. Figure 3.1 shows how a small change in the environment alters its beauty.

All these factors not only deteriorate nature but also have an adverse impact on human beings. Due to certain people's living circumstances and proximity to trash dumps, many suffer from ailments (Narayana, 2009). The pollution spreads like wildfire. No matter how great the challenges, there is always a window for improvement.

However, inadequate waste management exacerbates the issue, especially in developing countries and urban areas, highlighting the necessity of practical solutions to build sustainable societies (Malik et al., 2022). Trash must be categorized and segregated to reduce landfill space and combat the rise in global waste and greenhouse gas emissions, especially through recycling (Benjeddou et al., 2023). Even though more people are aware of the effects of climate change, many find it hard to change their habits. So, we really need solutions to help people separate waste (Singh, 2019). By recovering valuable materials, efficient waste categorization may improve recycling efforts in industries and thus result in cost savings. To prevent penalties and legal issues, it can also guarantee regulatory compliance. It can also reduce environmental impact, improving a company's sustainability status and public

DOI: 10.1201/9781003506478-3

FIGURE 3.1
Water pollution. (Source: Photo by Mohit Parashar: https://www.pexels.com/photo/photo-of-lake-during-daytime-3222575/)

impression. Lastly, it can reduce waste management costs and improve operational performance (Ao et al., 2022; Zhou et al., 2022).

The diversity of waste materials is vast, but the primary modules are paper, food waste, organic materials, and plastic. Due to these wastes, the manual handpicking of waste poses a high health risk for waste collectors. The waste management structure can be improved further by classifying this trash other than biodegradable and non-biodegradable. Electronic or sensor-equipped trash cans speed up the process (Ziouzios & Dasygenis, 2019). Classifying waste affects not just homes but also companies and enterprises. For effective waste management, then, smart trash classification methods are required.

The various nations of the world have come together to form certain alliances and have pledged to make the Earth a better place for future generations. Waste management generally refers to those activities that constitutes an object's origin to its demolition. The generation of waste annually is measured in tons, and its identification is a crucial topic for the implementation of systems that segregate it without impacting our environment. Everything comes with two parts: pros and cons. The amount of garbage produced and income are directly correlated. Figure 3.2 shows open waste disposal through which people earn their living.

FIGURE 3.2
Open waste disposal. (Source: Photo by Tom Fisk: https://www.pexels.com/photo/bird-s-eye-view-of-landfill-3181031/)

Waste has to be treated and disposed of as efficiently as possible, since it generates a significant amount of revenue. According to estimates, by 2050, the amount of waste produced daily will increase by 45% in poorer nations and 20% in affluent countries (Sharma & Jain, 2019).

With this type of experience, waste management services have been using new technology to better manage and recycle waste. This helps make waste sorting and recycling more efficient (Nowakowski & Pamuła, 2020).

Certain challenges are involved in the current waste management system:

- **Data Availability and Quality**: Reliable, current, and high-quality data are necessary for machine learning (ML) models to function. However, obtaining accurate and comprehensive data about waste generation and disposal can be challenging, particularly in developing countries or underprivileged areas.

- **Scalability and Computation**: The computationally intensive nature of training and evaluating machine learning models may limit their practical implementation in waste management systems. Solutions include optimizing the models' architecture and leveraging cloud computing or edge computing platforms.

- **Ethical Considerations**: Several ethical concerns might arise from the creation and implementation of a machine learning-based waste management system. For example, concerns about data privacy, biased algorithms, and potential harm to the environment from incorrect predictions may need to be addressed.

- **Time Constraints**: The efficiency of a machine learning-based waste management system may be impacted by time restrictions. It may take longer to develop, train, and implement such a system than it would with traditional methods, which could cause delays in resolving waste-related concerns.

Thus, in this work, we have investigated the efficacy of deep learning (DL) and machine learning models for waste categorization using image data. Utilizing the latest developments in computer vision and pattern recognition methods, our goal is to identify the optimal model for waste material classification that can distinguish between various types of waste materials by identifying visual signals that are derived from the image. The primary contribution of this paper is as follows:

- The Trashnet Dataset is used to train and evaluate these models to obtain a meaningful difference between machine learning and deep learning models in garbage classification.
- The purpose of data analysis is to have a comprehensive understanding of the data.
- Lastly, an analysis is made to determine how well each model performs in terms of accuracy, recall, and precision.

Figure 3.3 provides an overview of the proposed methodology and details the procedures that were followed to make the study a reality.

This is the format for the remainder of the paper. Section 3.2 contains the literature review. A description of the anticipated work is given in Section 3.3. Details regarding the suggested works' implementation and results are covered in Section 3.4, and Section 3.5 wraps up.

3.2 Related Work

By 2050, the world is expected to produce more than 27 billion tons of trash each year. China and India will be responsible for one-third of this waste (Majchrowska et al., 2022; Shahab et al., 2022). People who work with waste and don't have formal jobs deal with about 15%–20% of it. They might not be very knowledgeable about appropriate garbage management. Also, people may not be interested in reducing their waste or recycling it (Elkhouly et al., 2021). It might be difficult to integrate contemporary technology into the waste management process, especially when it comes to trash classification.

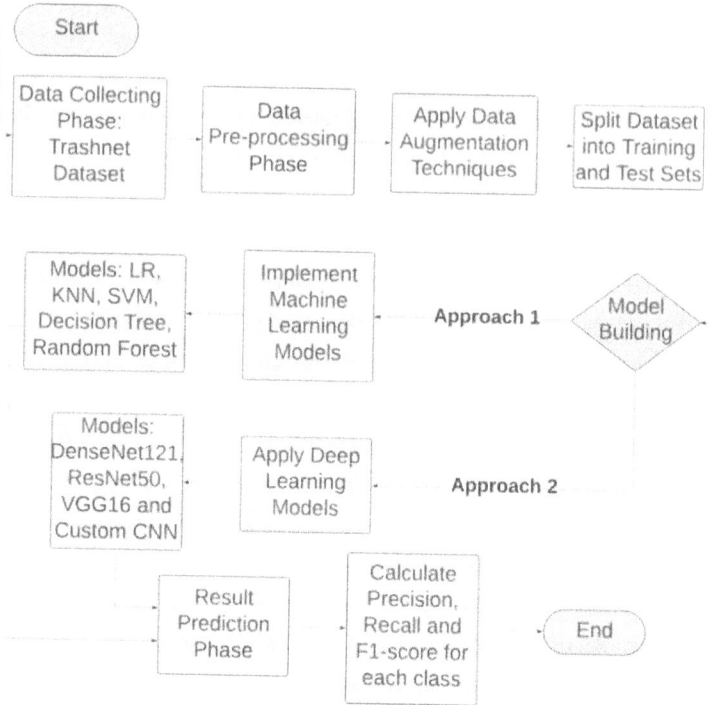

FIGURE 3.3
Block diagram of the proposed approach.

Trash sorting may be altered by computer vision and machine learning, which use technology to automatically recognize and separate various waste categories (Malik et al., 2022). By using technology to study pictures and train computers with machine learning models, it helps manage waste better and more efficiently, making efforts to protect the environment even stronger (Ao et al., 2022; Majchrowska et al., 2022; Shahab et al., 2022). This involves the usage of machine learning and deep learning-based techniques for waste cataloging. Numerous ways of using machine learning have been found in research to solve different problems (Bhatia & Rani, 2018), and the categorization of garbage is one such issue (Malik et al., 2022; Majchrowska et al., 2022; Shahab et al., 2022; Elkhouly et al., 2021; Bhatia & Rani, 2018; Rahman et al., 2022).

3.2.1 Machine Learning Approaches

In the realm of garbage categorization, (Yang et al., 2021) submitted GarbageNet, a learning framework that produced a mean accuracy of 87.69%. Mittal et al. (2016) suggested a phone app that made use of pictures taken by users to sort out different types of trash. Deep learning technology was used for this proposal. Spot Garbage employs Local Response Normalization with fully linked layers and CNN on images to identify trash. According to studies, CNN can recognize important traits without the assistance of a human (Acar et al., 2022). This means that CNN can function independently of humans. They are incredibly adept at classifying and recognizing images. Yang and Thung (2016) presented a system in 2016 for classifying garbage, namely trash, for the purpose of determining its recyclability status. They divided the waste into six categories. The public may access the dataset, which goes by the moniker Trashnet. When analysing the differences between several waste types, the authors used SVM and CNN. They found that SVM performs better than CNN. Their highest test accuracy of 84% led the developers to conclude that further dataset expansion is essential for the trash categorization task to be successfully completed.

Awe et al. (2017) classified trash using Trashnet and solved the same problem with a quicker R-CNN. The authors removed backgrounds from all images, blended multiple images, and then created three classes—"Landfill," "Paper," and "Recyclable"—and renamed them appropriately to ultimately achieve an accuracy of 68.3%. Faster R-CNN required more than 10,000 images, but Trashnet included about 2,500 images.

3.2.2 Computer Vision Approaches

In a different investigation, trash samples were gathered using a hyperspectral imaging framework system, and pre-processing was used to increase the precision of the categorization findings. To increase the accuracy of the classification results, the tests were put through rectification and denoising processes (Dong-e et al., 2019). Another model named GarbNet model was proposed by (Mikami et al., 2018), which attained an accuracy of 87.69%.

3.2.3 Ensembled Approaches

Several ensembled strategies were used in the literature, such as transfer learning. Using the idea of deep transfer learning and further improvements, (Vo et al., 2019) presented a garbage categorization system in 2019 that achieved a 94% accuracy rate across the Trashnet dataset. Some authors used the Attention-Knowledge component of the You-Only-Look-Once (YOLO) v5 model in 2021, which has the WSA attention component included as the spine of the YOLOv5 network for the classification purpose. For this task, the authors collected over 15,000 images of local garbage for model training, and the mAP achieved was 73.2% (Wu et al., 2021). A tool called SEFWaM (Goel et al., 2023) was presented for autonomous feature extraction from waste pictures. It was a CNN-XGBoost trained tool that attained 94.2% accuracy by utilizing the transfer learning method.

Table 3.1 shows a comprehensive review of the existing literature, citing their approach, advantages, and limitation(s).

TABLE 3.1

Comprehensive Review of the Existing Literature Citing Their Approach, Advantages and Limitation(s)

S.No.	Author & Year	Approach	Advantage(s)	Limitation(s)
1	Yang et al. (2021)	Proposed GarbageNet framework	Higher accuracy	Limited to a particular dataset
2	Mittal et al. (2016)	CNN-based waste classification on Spot garbage	Knowledge transfer	Data quality
3	Yang and Thung (2016)	Proposed method for waste recyclability status	Comprehensive approach	Real-time performance
4	Awe et al. (2017)	Faster R-CNN-based approach was proposed	Localization and data segmentation	Lower accuracy, data availability
5	Mikami et al. (2018)	Proposed a GarbNet model	Time efficient	Generalization
6	Vo et al. (2019)	Proposed a deep transfer learning-based approach	Knowledge transfer	Time complexity
7	Wu et al. (2021)	Approach used is YOLOv5 model	Combined dataset	Complexity, resource constraints

3.3 Proposed Work

We employed a comprehensive dataset for trash categorization in this work, which included pictures of several types of waste, including paper and plastic.

3.3.1 Data Gathering

GitHub provided the Trashnet2.0 dataset that was utilized in this study. This dataset consists of a directory containing six subfolders: cardboard, glass, paper, trash, metal, and plastic. These folders contain image data along with their respective labels.

The following Table 3.2 represents the data description:

Figure 3.4 illustrates some of the images that were picked by the model randomly.

TABLE 3.2

Trashnet Dataset Description

Classes	No. of Images
Cardboard	403
Glass	501
Metal	410
Paper	594
Plastic	482
Trash	137

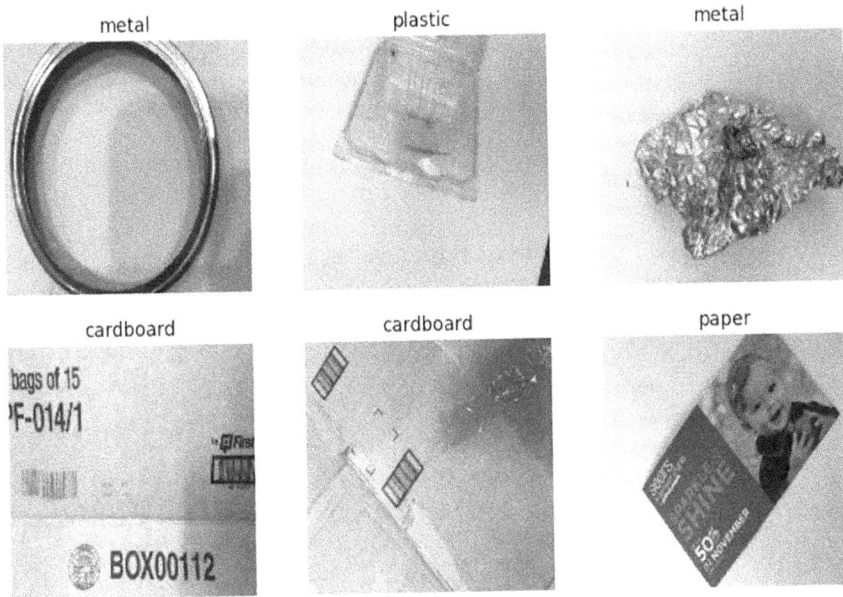

FIGURE 3.4
Random images from Trashnet dataset.

3.3.2 Pre-processing

For the purposes of training and testing models, the dataset is pre-processed to improve visual characteristics and standardize image resolution. The training and test sets of the dataset are split 80:20, respectively. In this phase, data augmentation techniques are applied to maximize data size, improve model generalization, and prevent overfitting. Section 3.3.2.1 discusses the augmentation techniques implemented for our work.

3.3.2.1 Data Augmentation

To enhance the variety of our training dataset, we employed data augmentation techniques, which are explained in this section. The implementation of augmentation strategies aimed to improve the resilience of our deep learning models, which in turn would reduce overfitting and thus help in model generalization.

- **Shear**: This transformation was introduced by up to 20%, as it introduces distortion in the image, mimicking some real-world deformations.

- **Zoom**: To simulate different distances between the camera and the objects, random zooming of up to 20% was added.

- **Rotation**: To improve visual replication of orientation, random rotation of up to [20–40] degrees was added.

- **Horizontal and Vertical Shifts**: To create spatial displacements, arbitrary shifts of up to 20% of the image's height and width were applied horizontally and vertically.

- **Horizontal and Vertical Flips**: Images were randomly rotated horizontally and vertically to introduce mirror reflections.

These augmented images were trained alongside the original images to improve the model's performance. Figure 3.5 shows the various augmentation techniques applied to one of the random images.

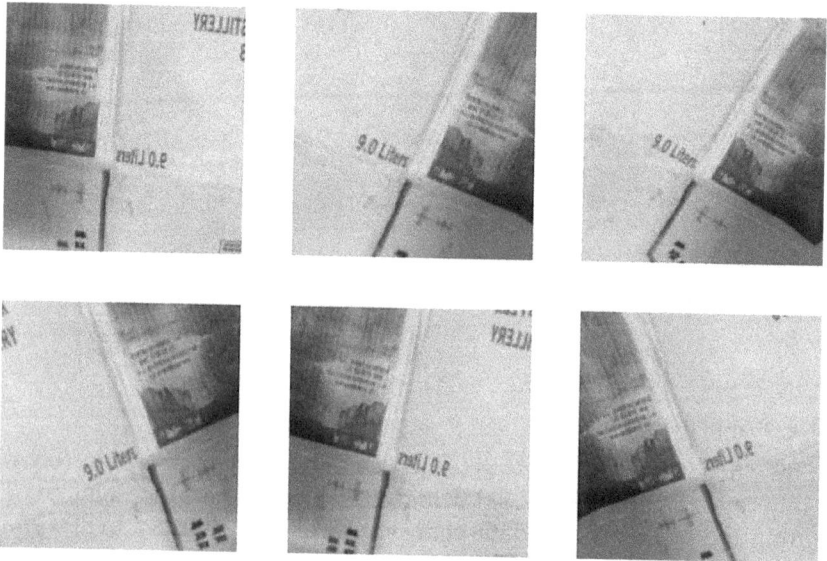

FIGURE 3.5
Various data augmentation techniques applied.

The ImageDataGenerator class from the TensorFlow/Keras package was used to create the data augmentation pipeline. The augmentation techniques and their corresponding transformation ranges were defined by utilizing the parameters of the ImageData-Generator class. During the model training process, the augmentation was done dynamically to guarantee that every batch of images fed into the models experienced random modifications. There was an 80%–20% ratio between training and validation data.

3.3.3 Model Building

Sections 3.3.3.1 and 3.3.3.2 describe the applied classification models for predictive analysis.

The models that are used are Decision Tree, Random Forest, SVM, k-Nearest Neighbor (KNN), and Logistic Regression (LR).

3.3.3.1 Machine Learning Models

A. Logistic Regression

A logistic regression model was used in the first run for the classification of trash into its respective categories.

With the use of a logistic function, this model converts a linear combination of input data into a probability.

$$P(y = 1/x) = 1/1 + e^{-z} \qquad (3.1)$$

where z is the input feature and the model coefficient linear combination.

The difference in log odds of the outcome for a single unit change in a predictor variable makes up the coefficients in logistic regression. This function quantifies the discrepancy between the actual class labels and the expected likelihood. We use the gradient descent technique to train our model.

B. Decision Tree

A decision tree represents a hierarchical structure comprising nodes that represent the features, branches that represent the decision rules, and leaves that represent the prediction. The splitting of the feature space at each node is on the basis of the information gain criterion. For prediction making, an instance traverses the tree from the root to a leaf node based on feature values.

C. Random Forest

During training, a random forest, which is part of the ensemble learning approach, creates a large number of decision trees and outputs the mode of the classes from each individual tree. At each decision tree node, the model chooses at random a subset of the features to be considered for splitting.

D. Support Vector Machine

The SVM was chosen as it is considered one of the best classification algorithms. It classifies items by defining a hyperplane that separates the data points into various classes. The SVM method searches the feature space for the decision boundary that best separates the classes. This boundary is defined by a subset of training samples, known as support vectors that are closest to the decision boundary.

E. K-Nearest Neighbor

KNN works by memorizing the entire training dataset and then makes predictions at runtime on the basis of similarity, using a majority vote between the new instances and existing data points.

3.3.3.2 Deep Learning Models

A. Convolutional Neural Network (Custom CNN)

We use the TensorFlow 2.9 framework for the construction of our own CNN. We implemented a twelve-layer CNN because of computational constraints.

- **Layer 0**: Input image of size 300x300
- **Layer 1**: Convolutional with 32 filters, size 3x3, stride 1, padding='same'
- **Layer 2**: Max-pooling with a pool size=2x2
- **Layer 3**: Convolutional with 64 filters, size 3x3, stride 1, padding='same'
- **Layer 4**: Max-pooling with a pool size=2x2
- **Layer 5**: Convolutional with 128 filters, size 3x3, stride 1, padding='same'
- **Layer 6**: Flatten
- **Layer 7**: Dense with 64 neurons
- **Layer 8**: Dropout rate of 0.2
- **Layer 9**: Dense with 32 neurons
- **Layer 10**: Dropout rate of 0.2
- **Layer 11**: Dense with 6 neurons

B. ResNet50

This study used the ResNet-50 model, a deep conventional neural network architecture with 50 layers of identity, pooling, convolutional, and fully connected blocks.

The utilization of ResNet-50 aligns with our objective of achieving high accuracy and robustness in waste classification. Its integration is leveraged by pre-trained weights obtained from public sources, as a pre-trained ResNet-50 model is fine-tuned using transfer learning techniques.

C. DenseNet121

As this is a comparison study, another deep convolutional neural network is taken into account, and that is the DenseNet-121 model. This model is renowned for its effective feature propagation and compactness.

The dense connectivity structure of the DenseNet-121 architecture is well-known, as every layer in a dense block receives direct inputs from every layer that came before it. Its integration is leveraged by pre-trained weights obtained from public sources as a pre-trained DenseNet-121model.

D. VGG-16

A different model, the VGG-16 model, is taken into consideration because of its efficacy and simplicity in the task of classifying images.

The defining characteristics of the 16-layer VGG-16 architecture are its depth and consistency, consisting of 13 convolutional layers and 3 fully connected layers. These layers use 3x3 convolutional filters to capture spatial information while maintaining a compact architecture.

3.3.4 Performance Evaluation

The models' performance is assessed in terms of precision, recall, accuracy, and F1 score. The results of the experimental evaluation are discussed in the following section.

3.4 Implementation and Results

Tensor-Flow supported the Keras framework, which was used to train the model. We compared and assessed the outcomes of multiple machine learning and deep learning models while accounting for different factors. To train the models, we utilized TensorFlow 2.9 on a Jupyter Notebook, with Python 2.9. Every deep learning model used an implementation of Adam's optimizer. Models were trained using the (model.fit) method on a pre-designated dataset. Using an iterative approach to minimize the loss function on the training data, this method carries out the real training procedure. One of the performance metrics that was put into practice was validation accuracy.

Our test results illustrate the comparative performance of ML and deep learning models for the waste classification tasks. Across the assessed measurements, deep learning models, especially CNNs, show prevalent execution compared to conventional ML models. The CNNs accomplished higher accuracy, recall, and precision in successfully recognizing between different types of waste materials on the basis of visual characteristics.

For the model's implementation, the system configuration required is shown in Table 3.3.

The ML models' performance is assessed based on F1 score, recall, accuracy, and precision as described in Table 3.4, Figures 3.6 and 3.7.

TABLE 3.3

System Configuration

Parameters	Configuration
Processor	Intel(R) core(TM)i5 2.50GHz
Memory	8GB
Operating System	Windows 10
GPU	12 GB

TABLE 3.4

Comparison of Various ML Models

Algorithm(s)	Accuracy
LR	0.47
KNN	0.44
SVM	0.65
Decision Tree	0.44
Random Forest	0.71

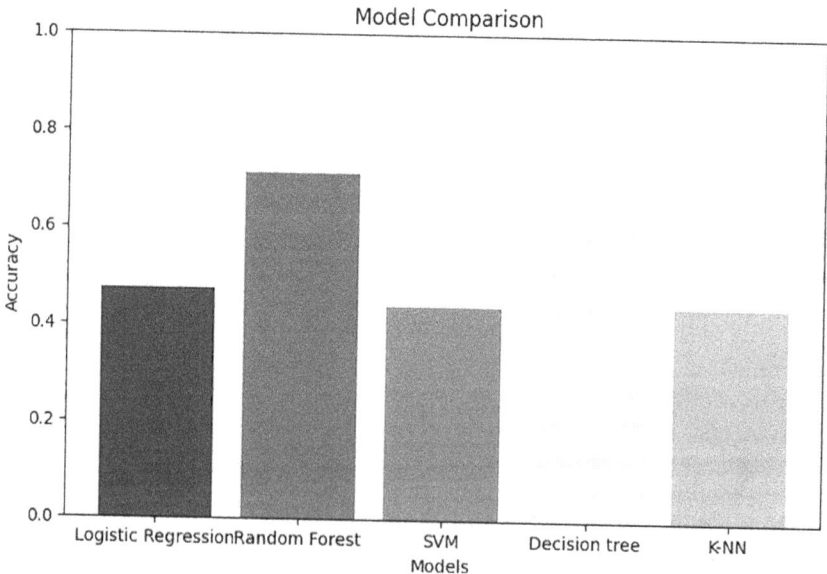

FIGURE 3.6

Accuracy comparison of ML models.

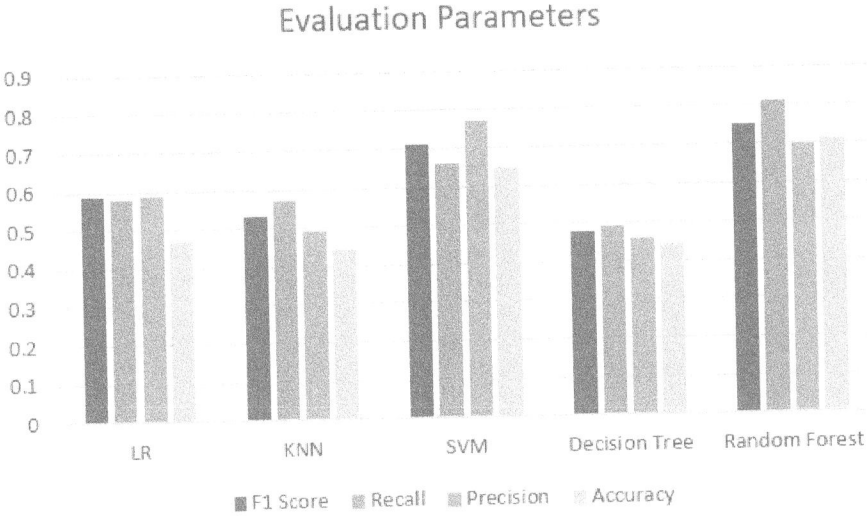

FIGURE 3.7
Evaluation parameters comparison of Trashnet dataset for ML models.

TABLE 3.5

Accuracy Comparison of Deep
Learning Models

Model	Accuracy
CNN	0.88
ResNet-50	0.42
DenseNet-121	0.95
VGG-16	0.66

Figure 3.6 displays the outcomes of the used machine learning algorithms. Random forest performs well in terms of accuracy. To avoid bias in this work, we employed fundamental methods.

For deep learning models, the graphical representation for comparison was not possible as their training was being done separately on different notebooks. However, their comparison is presented in Table 3.5.

The accuracy of CNN, ResNet-50, DenseNet-12, and VGG-16 is displayed in Table 3.5.

The other metrics, that is, recall, precision, and F1 score were calculated for each of the respective classes for both machine and deep learning models. Tables 3.6–3.9 represent the values for each of the classes in the dataset for the respective models.

TABLE 3.6

Metrics for VGG-16 Model

Categories	Precision	Recall	F1-Score
Cardboard	0.77	0.66	0.71
Glass	0.51	0.59	0.55
Metal	0.61	0.59	0.60
Paper	0.72	0.84	0.77
Plastic	0.66	0.62	0.64
Trash	0.82	0.31	0.45

TABLE 3.7

Metrics for ResNet-50 Model

Categories	Precision	Recall	F1 score
Cardboard	0.46	0.49	0.48
Glass	0.36	0.34	0.35
Metal	0.35	0.37	0.36
Paper	0.54	0.58	0.56
Plastic	0.52	0.40	0.45
Trash	0.26	0.31	0.29

TABLE 3.8

Metrics for Logistic Regression

Categories	Precision	Recall	F1 score
Cardboard	0.59	0.58	0.59
Glass	0.42	0.41	0.41
Metal	0.25	0.23	0.24
Paper	0.57	0.60	0.58
Plastic	0.47	0.51	0.49
Trash	0.39	0.38	0.39

TABLE 3.9

Metrics for Random Forest

Categories	Precision	Recall	F1 score
Cardboard	0.70	0.81	0.75
Glass	0.59	0.72	0.65
Metal	0.63	0.58	0.60
Paper	0.83	0.88	0.85
Plastic	0.78	0.60	0.68
Trash	0.85	0.38	0.52

3.5 Conclusion

Using a consistent dataset, this work provides a thorough comparison of machine learning (ML) and deep learning (DL) models, significantly contributing to the growing corpus of research on trash classification. The outcomes clearly demonstrate the superior and effectiveness of deep learning techniques—in particular, CNNs—at accurately identifying waste elements from visual input. These findings have significant implications for trash management in the future, especially in relation to automated garbage sorting, recycling practices, and environmental preservation initiatives. They also improve our understanding of waste classification.

In the future, more studies are needed to investigate multi-modal data fusion methods, transfer learning strategies, and sophisticated deep learning architectures. These types of studies will be essential in clearing the path for reliable and expandable waste management systems. Furthermore, assuring the efficacy and dependability of the next waste management systems will require tackling the issues related to data scarcity and the scalability of waste classification databases. By concentrating on these areas, we can support waste management and environmental preservation strategies that are more effective and sustainable.

References

Acar, Z. Y., Başçiftçi, F., & Ekmekci, A. H. (2022). A Convolutional Neural Network model for identifying Multiple Sclerosis on brain FLAIR MRI. *Sustainable Computing: Informatics and Systems*, 35, 100706. https://doi.org/10.1016/j.suscom.2022.100706

Ao, Y., Zhu, H., Wang, Y., Zhang, J., & Chang, Y. (2022). Identifying the driving factors of rural residents' household waste classification behavior: Evidence from Sichuan, China. *Resources, Conservation and Recycling*, 180, 106159. https://doi.org/10.1016/j.resconrec.2022.106159

Awe, O., Mengistu, R., & Sreedhar, V. (2017). Smart trash net: Waste localization and classification. arXiv preprint. https://api.semanticscholar.org/CorpusID:36189518

Benjeddou, O., Ravindran, G., & Abdelzaher, M. A. (2023). Thermal and acoustic features of lightweight concrete based on marble wastes and expanded perlite aggregate. *Buildings*, 13(4), 992. https://doi.org/10.3390/buildings13040992

Bhatia, V., & Rani, R. (2018). Dfuzzy: A deep learning-based fuzzy clustering model for large graphs. *Knowledge and Information Systems*, 57, 159–181. https://doi.org/10.1007/s10115-018-1156-3

Dong-e, Z., Rui, W. U., Bao-guo, Z., & Yuan-yuan, C. (2019). Research on garbage classification and recognition based on hyperspectral imaging technology. *Spectroscopy and Spectral Analysis*, 39(3), 917–922. https://doi.org/10.1109/ITAIC54216.2022.9836699

Elkhouly, H. I., Abdelzaher, M. A., & El-Kattan, I. M. (2021). Experimental and modeling investigation of physicomechanical properties and firing resistivity of cement pastes incorporation of micro-date seed waste. *Iranian Journal of Science and Technology, Transactions of Civil Engineering*, 1–13. https://doi.org/10.1007/s40996-021-00760-2

Goel, S., Mishra, A., Dua, G., & Bhatia, V. (2023). SEFWaM–deep learning based smart ensembled framework for waste management. *Environment, Development and Sustainability*, 1–29. https://doi.org/10.1007/s10668-023-03568-4

Majchrowska, S., Mikołajczyk, A., Ferlin, M., Klawikowska, Z., Plantykow, M. A., Kwasigroch, A., & Majek, K. (2022). Deep learning-based waste detection in natural and urban environments. *Waste Management*, 138, 274–284. https://doi.org/10.1016/j.wasman.2021.12.001

Malik, M., Sharma, S., Uddin, M., Chen, C. L., Wu, C. M., Soni, P., & Chaudhary, S. (2022). Waste classification for sustainable development using image recognition with deep learning neural network models. *Sustainability*, 14(12), 7222. https://doi.org/10.3390/su14127222

Mikami, K., Chen, Y., Nakazawa, J., Iida, Y., Kishimoto, Y., & Oya, Y. (2018, August). Deepcounter: Using deep learning to count garbage bags. In *2018 IEEE 24th International Conference on Embedded and Real-Time Computing Systems and Applications (RTCSA)* (pp. 1–10). IEEE. https://doi.org/10.1109/RTCSA.2018.00010

Mittal, G., Yagnik, K. B., Garg, M., & Krishnan, N. C. (2016, September). Spotgarbage: Smartphone app to detect garbage using deep learning. In *Proceedings of the 2016 ACM International Joint Conference on Pervasive and Ubiquitous Computing* (pp. 940–945). https://doi.org/10.1145/2971648.2971731

Narayana, T. (2009). Municipal solid waste management in India: From waste disposal to recovery of resources? *Waste Management*, 29(3), 1163–1166. https://doi.org/10.1016/j.wasman.2008.06.038

Nowakowski, P., & Pamuła, T. (2020). Application of deep learning object classifier to improve e-waste collection planning. *Waste Management*, 109, 1–9. https://doi.org/10.1016/j.wasman.2020.04.041

Rahman, M. W., Islam, R., Hasan, A., Bithi, N. I., Hasan, M. M., & Rahman, M. M. (2022). Intelligent waste management system using deep learning with IoT. *Journal of King Saud University-Computer and Information Sciences*, 34(5), 2072–2087. https://doi.org/10.1016/j.jksuci.2020.08.016

Shahab, S., Anjum, M., & Umar, M. S. (2022). Deep learning applications in solid waste management: A deep literature review. *International Journal of Advanced Computer Science and Applications*, 13(3). https://doi.org/10.14569/ijacsa.2022.0130347

Sharma, K. D., & Jain, S. (2019). Overview of municipal solid waste generation, composition, and management in India. *Journal of Environmental Engineering*, 145(3), 04018143. https://doi.org/10.1061/(ASCE)EE.1943-7870.0001490

Singh, A. (2019). Managing the uncertainty problems of municipal solid waste disposal. *Journal of Environmental Management, 240,* 259–265. https://doi.org/10.1016/j.jenvman.2019.03.025

Vo, A. H., Vo, M. T., & Le, T. (2019). A novel framework for trash classification using deep transfer learning. *IEEE Access, 7,* 178631–178639. https://doi.org/10.1109/ACCESS.2019.2959033

Wu, Y., Shen, X., Liu, Q., Xiao, F., & Li, C. (2021). A garbage detection and classification method based on visual scene understanding in the home environment. *Complexity, 2021,* 1–14. https://doi.org/10.1155/2021/1055604

Yang, J., Zeng, Z., Wang, K., Zou, H., & Xie, L. (2021). GarbageNet: A unified learning framework for robust garbage classification. *IEEE Transactions on Artificial Intelligence, 2*(4), 372–380. https://doi.org/10.1109/TAI.2021.3081055

Yang, M., & Thung, G. (2016). Classification of trash for recyclability status. *CS229 Project Report, 2016*(1), 3. https://api.semanticscholar.org/CorpusID:27517432

Zhou, H., Yu, X., Alhaskawi, A., Dong, Y., Wang, Z., Jin, Q., ... & Lu, H. (2022). A deep learning approach for medical waste classification. *Scientific Reports, 12*(1), 2159. https://doi.org/10.1038/s41598-022-06146-2

Ziouzios, D., & Dasygenis, M. (2019). A smart recycling bin for waste classification. In *2019 Panhellenic Conference on Electronics & Telecommunications (PACET)* (pp. 1–4). IEEE. https://doi.org/10.1109/PACET48583.2019.8956270

4

AI-Powered Data Analytics for Optimizing Tea Tourism: An Investigation on Determinants and Destination Information

Shuvasree Banerjee, Mohammad Badruddoza Talukder, and Viana Hassan

4.1 Overview

4.1.1 Context and Significance

There is an increasing consensus among individuals on what constitutes being beneficial or detrimental for their health and welfare, attributed to a proactive improvement in living. The consumption of tea is swiftly increasing in tandem with advancements in network technology and a rise in information absorptive capacity. Consequently, the overall number of individuals acquiring knowledge about the essential components included in tea is increasing. Individuals consistently embrace the consumption of tea and associated products (Banerjee & Tyagi 2024). A transformation in consumption patterns is seen due to economic expansion and rising affluence. It is contended that simply possessing sufficient material goods to satisfy the fundamentally defined wants of individuals is insufficient for a substantial consumer population today. The significance of spiritual culture has been progressively increasing as more individuals become familiar with it. A rising tendency has emerged in which individuals engage in spiritual activities to evade societal pressure (Bhatia 2006). Individuals may pursue specific objectives such as weight reduction, career advancement, or enhancement of skills in areas such as music or dance. Participation in tourism activities would undoubtedly benefit the vast majority of individuals. Environmental tourism, sometimes referred to as green tourism, occurs when tea estates and tourism enterprises collaborate. The integration of AI in tea garden ecotourism can facilitate the sustainable expansion of the tourism sector. A novel method, including ecotourism, is being implemented within the tea business. It can promote economic growth, enhance the popularity of tea culture in the area, and support the sustainable development of the Indian tourism

DOI: 10.1201/9781003506478-4

industry. At the same time, these can promote mutual development of local ecosystems, communities, and enterprises (Bora & Bora 2005). This topic is related to the identification of tea garden ecotourism and the support of its sustainable development. It is a social obligation that should be ensured.

4.2 Organic Agriculture's Ability to Facilitate Sustainable Ecotourism in Tea Gardens

4.2.1 Semantic Analysis of the Multifunctional Concept of Organic Farming

4.2.1.1 The Significance of Organic Farming's Diverse Advantages

Agriculture is the cultivation of arable land, water, and animals for food or by-products. One of the positive and eco-friendly trends in modern development is the transition from conventional agriculture to organic farming. Organic farmers use only natural fertilizers, while conventional farmers use several chemical fertilizers, herbicides, and feed to increase plant and animal productivity and crops Das (2009). The special dietary needs of plants and animals are beneficial for humanity and the world. Agricultural development will be beneficial for the economy, ecology, customs, and population, as well as for other areas. The main ecological advantage of organic agriculture highlighted in the article is flexibility. A consideration of relevant concepts on agricultural flexibility provides the following definition of adaptation in organic farming: the ability to use organic farming methods provides the potential for the development of the population and the land supporting life. Das (2008) wrote, "National political stability, economic growth, and social development are all beneficial and contribute significantly." The organism's system is realized when one of the subfunctions affects and is connected with another.

4.2.1.2 Constituents of the Multifunctionality of Organic Agriculture

Organic farming is the process of facilitating food production. The previous section illustrates that organic farming has a lot of flexibility due to three factors. The environment is the primary concern regarding the impact of agricultural activities. Geography, climate, hydrology, and other biophysical elements may also be referred to as resources. Many individuals strictly advocate for organic farming due to its use of accessible natural resources toward food production and, therefore, its eco-friendliness (Dutta et al., 2020). Changes in terrain and the availability of resources can provide alternative views for landscapes, which have served many functions. These objectives are directed to tourists, income, and locations for conservation. These are new social functioning determinants, such as environmental carrying capacity and human health concerns, among others. Diversity influences individuals' attitudes and cognition, therefore acting as a medium for cultural education. Alternatively, one can encircle a natural source of food. Organic farming ensures farmers' protection during farming activities, raises the yield of crops, raises farmers' income, and minimizes soil nutrient loss due to excessive use of fertilizers and pest control measures. It also offers different agricultural and allied products that are less harmful to people and the environment. Organic cultural development seems to be an indirect outcome of extensive modernization processes. Analysis of culture could enhance the comprehension of society's norms and values. Lastly, organic agriculture and by-products could satisfy customers' nutritional and health needs in rural and urban regions, as well as provide raw materials for industrial companies. The social and political life within a country is defined by the nature of food, including organic farming and its produce. Tertiary and farm products can hence be said to be the end products of synthetic farming (Goowalla, 2012). Contemporary cultural and historical significance of characteristics inherited from the past, fashioned by natural selection, remains extremely strong.

4.2.1.3 Dimension of Organic Agriculture

Organic agriculture is a holistic concept, and the diversity involved is mainly based on five important facets:
The economic function owes its origin primarily to the role of its output in contributing to the socio-economic viability of society, particularly in the context of organic agriculture and related sectors.

- **Social Function**: The quality of organic agricultural by-products' greatly affects human health and social welfare, since they are related to essential issues in society such as unemployment and underemployment; therefore, organic agricultural resources dictate the quality of these factors.

- **Political Role**: It is imperative to recognize that agriculture is the most important sector of our country's economy. It should be noted that the country has had primary dependence on agriculture for thousands of years. It enforces law and order in society, demarcates social stratification, and determines the government and economic position of a country.

- **Ecological Function**: Ecological purposes are served by organic agriculture, as its components are intrinsic aspects of the ecological environment.

- **Cultural Function**: Agriculture is a sacred profession inextricably associated with ancient civilization. People gain new knowledge and change their point of view regarding art and culture.

4.2.2 Ecotourism in Tea Gardens

4.2.2.1 Definition of Ecotourism in Tea Gardens

The Tea Garden Ecotourism refers to the combination of nature tourism and herbal plant agriculture tourism, directly related to ecological conservation. To successfully establish ecotourism, there is a need to develop tea plantations and use tea gardens properly to fit within natural, cultural, and modern tourist environments. Therefore, it is a new type of ecotourism focusing not just on the maintenance of existing tourist facilities in rural areas but also on improving them (Goowalla & Neog, 2011). The former is used to refer to the land allocated for tea cultivation, whereas the latter is used to refer to the place where tea products are consumed. The former is characterized as a tea garden in this case. Tea gardens are dependent on tourist development due to their beautiful landscapes and great aesthetic value caused by environmental differences. The term for excursions to tourist places covering a visit to a tea garden is referred to as tea garden ecotourism trips.

4.2.2.2 Present status of Tea Garden Ecotourism

It can also arise from the idea that people use holidays as a time to recharge themselves physically and mentally; as they become wiser and more experienced, they tend to see holidays as an investment in their growth and health rather than in the accumulation of commodities. Tourism, environmental immersion, and taking care of one's physical and spiritual health are considered necessary by the majority, at least according to the survey, for a complete life. Tea garden tourism is important to the development of the industry and will help create awareness of the topic (Hazarika 2011).

However, claiming that the significant amount of effort put into tea garden ecotourism was not suitably rewarded would be unfair; nevertheless, the conventional model of tea garden ecotourism is now obsolete because social and customer tastes change. In environments where there is clarity, people possess an enhanced ability to doubt such matters. The tea garden ecotourism area is plagued by poor infrastructure; on the other hand, in areas of high population density, the high demand for accommodation in the peak season creates a shortage of available accommodations and restaurants for travelers. Following cars too closely influences normal traffic jams, and the observation area only provides one type of tea to its customers. The importance of clean, well-preserved tourist destinations is evident: trash can completely take away from what would otherwise be a pleasant visit for tourists (Hazarika & Muraleedharan, 2011). The tea garden ecotourism industry has major opportunities; however, it also encounters several challenges. The best method to carry out an overall analysis of the challenges affecting tea garden ecotourism is to develop well-defined responses to possible difficulties, hence the need to provide treatments with sufficient caution to ensure long-term sustainability.

4.2.2.3 Tea Garden Ecotourism Development Criteria

In the course of developing ecotourism in tea gardens, the following concepts and standards need to be followed. Creating a tea garden ecotourism model requires the following ideas and requirements to be followed:

- **Reality-based and Customer Demand Fulfilment-Oriented**: This Mobile Information System needs to consider all consumer needs and ensure that this market segment does not confuse competitors by matching exactly with those needs.

- **In-depth Scrutiny of a Topic by Topic**: As one advances in the model of ecotourism in tea gardens, it is imperative to come up with solutions rather than copying examples from other gardens. Being aware of other organizations' progress requires a focus on highlighting an equally strong and innovative new development.

- **Have an Overall View While Taking into Account the Development of Tea Gardens**: The functioning of tea gardens as part of ecotourism activities involves activities related to the construction of new beneficial facilities for tourists, such as tea gardens, along with the continued upgrading and upkeep of existing amenities. When picking a site for your tea garden, it is essential to examine the clientele interested in the products and services provided by the business, and one should strive to provide customer happiness in every conceivable manner. The negative impacts necessarily increase in terms of the economy, environment, and social aspects of any country. Policy actions must be taken to ensure that the development and advancement of tea garden ecotourism do not have negative impacts on the biophysical environment. It is argued that the institutions of ecotourism in tea gardens deserve fair attention and motivation in both developmental and conservation aspects.

4.2.3 Approaches to Sustainable Ecotourism Development in Multifunctional Organic Tea Estates

4.2.3.1 Agriculture

Organic farming has been marked by its political, cultural, social, ecological, and economic advantages. Tea estates are required under the agricultural practice of tea cultivation. Tourist landscapes of outstanding visual quality can be formed because of the varied topography, climate, and other natural features of the region. Tea garden ecotourism has to quickly embrace a new model of development based on the flexibility of organic agriculture so that consumer needs cannot be met by the conventional tourist scene of tea gardens (Hazarika & Muraleedharan, 2011). Proper utilization of the economic value of tea gardens, along with their environmental and educational value through responsible ecotourism, is imperative for enhancing sustainable development. Improve the experience of visitors through the development of the tea garden's' existing infrastructure. The area remains largely undeveloped, despite the success of the tea garden tourism industry.

The majority of tea farms do not have unique selling propositions in management and tourism-related attractions. Secondly, there is a huge lack of support infrastructure for tea gardens. The limited availability of accessible routes of traffic and hospitality facilities significantly impacts the number of visitors to the gardens; therefore, sustainable tea garden ecotourism development should start with infrastructure development to realize the potential of the garden. Additionally, the existing brand of the tea garden will limit its expansion. Therefore, it is necessary to maximize the resources of the garden, develop distinctive tourist products, and link social and environmental resources to realize the' full potential of tea garden ecotourism. To attract more tourists, the garden's' infrastructure and supporting facilities must be enhanced. A development model should be based on the tea industry, with other industries acting as sub-models, while emphasizing tea plantations and tea cultivators (Hussain & Hazarika, 2010). Increasing the number of innovative and exhaustive activities dedicated to tea culture tourism could lead to sustainable growth in tea garden ecotourism. Tea is the sole product of a tea garden, and the architecture of the garden should be designed in a way that does not hinder the view of the tea leaves. The evolution of tea garden ecotourism must emphasize tea production and cultivation. Improved tea output through standardized administration and cultivation of tea bushes can be achieved, for example, by increasing investment in the education and training of tea farmers and other technical and professional staff (ITA, 2009). Through the creation of multiple resources related to tea, the tea garden can be expanded to include activities such as tea farming, tea picking, and travel. The many benefits and applications of tea gardens should guide the creation of ecotourism in these regions. Having access to an extensive range of material sources makes it possible to meet a wide range of client' needs. Achieving equilibrium between enhancement and safeguarding, between fundamental principles and commercial platforms, and between essential needs and regulatory frameworks is crucial. To ensure the sustainability of ecotourism in tea gardens, it is essential to consider the multifunctional requirements of organic agriculture during the planning process (Jolliffe, 2007). Development and environmental conservation must be given equal importance, as should the consumer market and the tourism sector, alongside national policy and consumer demand. It is vital to consider both economic and ecological benefits when planning tea gardens for ecotourism. Tea gardens must be designed to fit consumer consumption patterns by incorporating appropriate cultural and environmental elements into Mobile Information Systems (Jolliffe, 2010), rather than developing without these factors in mind. Lastly, developing tea gardens into ecotourism resorts can be costly; therefore, it is crucial to make use of available government aid programs.

4.3 Artificial Intelligence Technologies for choosing Tea Destinations

A thrilling prospect to enrich the visitor experience through personalized recommendations has arisen by integrating AI and sophisticated algorithms into tea tourism practices. These models apply machine learning and data analysis to match visitor interests with suitable activities and destinations. To deliver personalized recommendations, recommendation systems within the tea tourism sector employ algorithms to interpret user activities and interests. They take into consideration data such as search history, demographics, and past travel experiences to create personalized recommendations (Kim & Park., 2013).

Destinations can be matched by content-based filtering, where factors such as cultural events, historic landmarks, or tea plantations are considered, while collaborative filtering uses user interests to offer recommendations. Sentiment and emotional tone of text content, such as social media posts, reviews, and comments, can be measured through Natural Language Processing (NLP) methods. This sentiment analysis plays an important role in enhancing the perception of various tea tourism destinations held by the general public. Travelers looking for tea can apply picture recognition technology to assist them in recognizing and categorizing landscapes, cultural festivals, and famous tea destinations, said Lee Jolliffe in 2007. By responding to questions, assisting in trip planning, and providing live support, chatbots and artificial intelligence-based virtual assistants enhance user engagement. Lastly, AI-based tea tourism systems provide numerous services that enrich the entire experience for visitors. Travel can be enhanced to be more interesting, interactive, and tailored with the incorporation of extensive data analysis, image identification, natural language processing, and live user interaction (UX).

4.4 Travel Recommendation Artificial Intelligence Platforms: Limited Application

a. TripAdvisor

The renowned travel website TripAdvisor employs artificial intelligence and machine learning software to recommend hotels, restaurants, tourist sites, and other travel destinations according to individual consumer preferences. TripAdvisor aims to leverage users' ratings, reviews, and likes through technology to provide travelers with customized recommendations, making them better-informed decision-makers.

b. Expedia

If you want a holiday package that suits your preferences and stays within your budget, Expedia will assist you in finding it using artificial intelligence algorithms. Site recommendations and search results are constantly improved through machine learning algorithms.

c. Skyscanner

Utilizing artificial intelligence, the Skyscanner travel search engine finds the best prices, the most accommodating dates, and the lowest fares. As per historical data, it predicts the optimal time to book flights. One such app is Hopper, which employs AI to forecast travel expenses such as plane fares and hotel stay. Customers can make informed choices and potentially save money by following the software's advice on whether to purchase immediately or wait for better fares.

d. Google Trips

Google Flights and Google Maps are two of the numerous travel products that constitute Google Travel. In an attempt to give users personalized vacation suggestions, the sites utilize artificial intelligence. Google Flights makes predictions about price shifts with machine intelligence, while Google Maps directs you and provides destinations to visit in real time.

4.5 An Analysis of India's Tea Market

Stretching over 312,210 hectares, tea gardens in Assam produce 507 million metric tons of tea annually. The majestic Brahmaputra River flows through the region, and the one-horned rhino resides here. Annual rainfall is between 2,500 and 3,000 mm, and the altitude ranges between 45 and 60 m above sea level. The teas of the region have a special niche in the market, as Assam receives 100–150 inches of rain annually (Figure 4.1).

4.5.1 Assam

Assam teas are famous for their strong taste and fragrant aroma; the name says it all about their fame. Assam is "your cup of tea" if you want strong, tasty tea.

4.5.2 Darjeeling

At 17,820 hectares, you can locate the tea plantation. Results: 9.8 million kg, 90–1,750 m elevation. Total precipitation: 3,300–3,600 mm. Standing at 600–2,000 m behind the snow-capped Himalayas, Darjeeling produces one of the

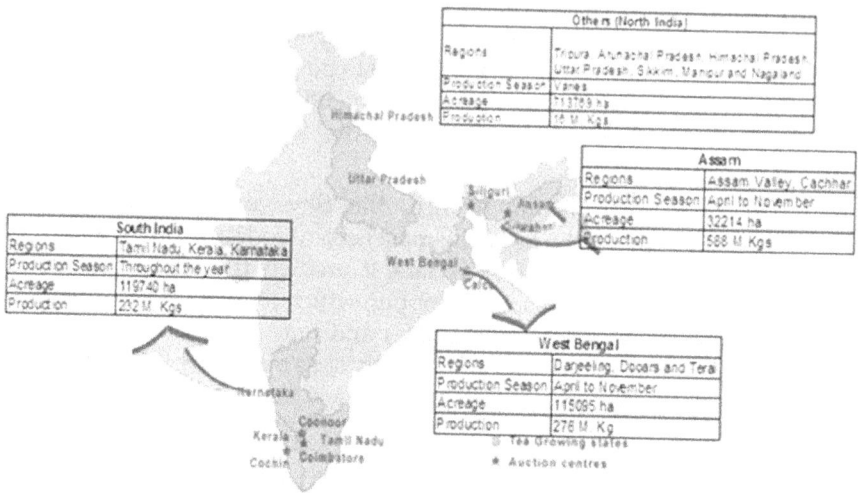

FIGURE 4.1

Tea production map in India. (Source: https://www.indiatea.org/)

finest teas in the world. Darjeeling tea tends to be green or gold in color and also green or gold in flavor and can be defined as "muscatel," "flowery," "peachy," or "tasting like muscatel grapes." Since its flavor is mild and agreeable, it is preferred by most individuals to be consumed unsweetened, without milk and sugar, rather than the darker, bitter teas. The idea of Darjeeling tea faking leaves specialists with a frown. Landscape and herds, the whole tea-growing area covers an elevation range of 90–1,750 m, covering 97,280 acres. With rainfall of 3,500 mm, by 1874 the Darjeeling district had a whopping 113 tea gardens. It is for this reason that the Terai farmers started cultivating tea. In 1862, the Terai farmers started cultivating tea because of Landscape and herds.

4.5.3 Dooars

White set up Champta, the first plantation in the Terai district. The Dooars were included in the garden. However, the British established the Dooars Tea Planters' Association in 1877 as a result of the better performance of the Assamese tea bush in this area. Thirteen plantations had developed in the area in 1876, with Gazeldubi being the initial Dooars garden. About a quarter of India's yearly tea production is supplied by the Dooars and Terai gardens, with output hitting 226 million kg in September. There are 48 gardens in the TAI belt. The Dooars are an incredibly lovely place located in the Himalayan foothills. Green forests sustain a rich variety of wildlife, and the landscape ranges from gently sloping plains to low hills rising out of the rivers. "Dooars" means "doors" because they act as a gateway between Bhutan and northeast India. The Dooars have been inviting tourists to the hill resort town of Darjeeling and the Sikkim region with their world-renowned tea gardens, which were initially set up by the British. The altitude range in the Dooars region is 150–1,750 m. The mountains of Bhutan are the source of numerous rivers and streams that nourish these fertile lowlands. About 3,500 mm of precipitation fall each year on average. Around mid-May is usually when the monsoon season starts, lasting throughout the entire month of September. Regular winter weather features chilly temperatures and foggy mornings and evenings. The summer is short and moderate, lasting only a couple of months. The economy is fueled by the "three Ts": tea, tourism, and timber. Tens of thousands of people work as manual laborers in the tea industry and neighboring companies.

4.5.4 Kangra

Maximum output: 0.8 million kg. The planted area covers 2,348 hectares. Conditions: 700–1,000 m in altitude, 2,300–2,500 mm of precipitation. The districts of Mandi and Kangra in Himachal Pradesh are home to 2,063 acres of tea plantations. The unique taste of Kangra tea—commonly referred to as coming from "the valley of gods"—has made it renowned throughout the globe. The rolling hills of the outer Himalayas, under the awe-inspiring snow-covered Dhauladhar Mountains, have been tea-growing lands since 1949.

4.5.5 Mount Nilgiri

Our tea garden covers 66,175 hectares, producing 135,000,000 kg of tea annually. The region receives 1,000–1,500 mm of rainfall and lies at an altitude of 1,050–2,634 m above sea level. The Nilgiris, or the Blue Mountains, are the homeland of Nilgiri tea, which is among India's finest. The Blue Mountains were named after the sax-blue Kurinji flower, which blooms only once every 12 years. Tea is cultivated at elevations ranging from 1,000–2,500 m in this beautiful part of southern India's rolling hills. It can rain anywhere from 60 to 90 inches annually. These conditions are best suited for Nilgiri teas because they enhance their strong, sophisticated flavor and bright infusion.

The unmatched aroma and stimulating taste of Nilgiri distinguish it from other teas. This is the perfect tea for those looking for something strong, savory, and scented. Tea characteristics: a scented, floral tea with an intense floral bouquet and a sunny yellow appearance. Light and very brisk. Invigorating fragrances of late-blooming flowers linger in the atmosphere. The mouthfeel is like cream. Ideal for relaxing after a tiring day. Nilgiri Orthodox Tea is a recently registered GI by the Indian government.

Figure 4.2 indicates that India has experienced numerous fluctuations in tea production over the years. The tea crop during 2017–2018 was 1,325.1 million kg, as compared to 1,344.4 million kg in the year 2016–2017. Manufactured tea is on the rise, though at a gradual pace. Domestic tea manufacturing as well as exports are on the rise.

Tea production in India (million kg)

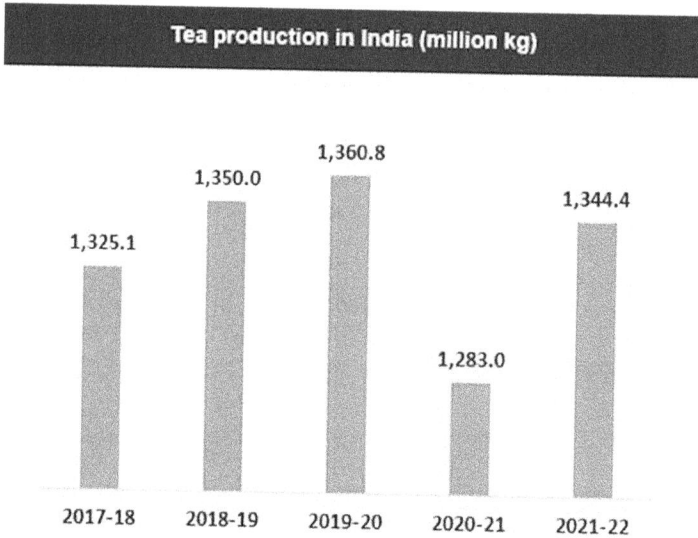

Source: Tea Board of India

FIGURE 4.2
Tea production in India.

4.6 Positive Impacts of AI on Tea Tourism and Economic Development

- **Personalized Experiences to Suit Your Needs**: Through AI-based recommendation systems, tea tourists can have their journeys customized to particular estates, occasions, cultural activities, and local sites that suit their specific interests in tea. Personalized travel plans ensure that tourists are more likely to have a great time, give good feedback, and recommend places to others.

- To discover the best tea shops, marketing and engagement tools with AI capabilities can analyze user reviews and ratings. Such insights can enable businesses and tourism boards to focus their advertising more effectively.

- **Artificial intelligence applications** in tea-growing areas could assist in promoting improved sustainable tourist development by minimizing issues such as overpopulation and ecological degradation. Policies of sustainable tourism could be improved and implemented to guard the environment and cultural heritage through artificial intelligence technologies, enabling historical research and tourist trend studies.

- **Empowerment and Community Involvement**: Utilization of AI-powered platforms can ensure empowerment and community involvement in the emerging tea tourism sector. The application of chatbots and virtual assistants has excellent potential for enhancing local enterprises' and craftsmen's revenues, along with providing better opportunities for cultural immersion for visitors.

- **AI analytics** can provide useful information on visitor preferences, dislikes, and expenditures, enabling players in the tea tourism sector to make evidence-based choices. Such evidence-based understanding of what attracts visitors, what the effective drawcards are, and where to put infrastructure investments may inform strategic choices.

- **Localized Recommendations:** With the use of AI to find less popular or undiscovered treasure spots within tea-growing areas, the benefits of tourism can be brought to less-popular areas. Due to this, the local economy can grow more evenly, and tourist congestion at top destinations will be reduced.

- **Virtual assistants or chatbots** based on artificial intelligence can offer tourists real-time support, from answering questions about local traditions and public transport options to providing information on tea-related activities. This improves the general experience for tourists and encourages good reviews, which can lead to an influx of visitors.

Artificial intelligence programs can process past and present information to forecast the maximum demand for tea tourism. Local businesses will be able to prepare for any contingency, collect materials, and track employee locations using this data.

AI may help the tea tourism industry provide better customer care by analyzing tourists' reviews and feedback. The government and local entrepreneurs can always use feedback to improve things and solve any problems that arise.

4.7 India's Tea Tourism

Visiting India to tour the nation's tea-growing districts has become a favorite activity among tea connoisseurs, who are now able to fully experience India's vast culture and stunning scenery. Tea tourism, an interesting and innovative way of traveling in India, is now very much in vogue. The tea from the Nilgiri Hills, Darjeeling, and Assam is a favorite among Indian tea drinkers because of the perfect hilly terrain and large tea gardens located here. From the precise oxidation and drying to the manual selection of young shoots, tourists see it all on tea tours. They can learn more about the intricate tea-growing process by joining expert guides on tours of breathtaking estates, where they may also encounter local tea pluckers. Overnight stays are possible in a quaint colonial home or a tea cottage on the estate, and the life of a tea producer can be experienced if desired. The tours won't be complete without tea-tasting sessions, which are usually conducted in locations with scenic beauty. People interested in tea, nature, and serene getaways prefer reserving this sort of getaway (Meshram et al., 2022). Set forth on an interesting tea tourism experience that combines the rich cultural heritage, dramatic natural beauty, and centuries-old tea culture of India. Tourists participate in the process as they walk through the tea gardens. Tourists worldwide will be captivated by the assiduous history of Assam tea, which is responsible for the unique flavor and aroma. Increasingly, travelers from around the world are keen to visit the Assam Tea Estates and discover different aspects of Assam tea (see Figure 4.3).

Assam tea tourism can generate a great deal of revenue for the state due to its lovely gardens, the so-called "Green Carpet" with trees covering it, the tea estate cottages reminding one of the state's British colonial heritage, and the tanned ethnic locals with their own distinct music and culture. It is a potentially very profitable and financially successful venture for this area. The growing popularity of tea tourism has led to dramatic shifts in South Asian countries such as India's travel markets (Figures 4.4 and 4.5).

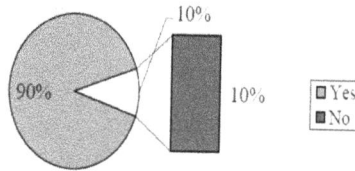

FIGURE 4.3
Tea tours help in promoting tourism in Assam. (Source: Self adaptation.)

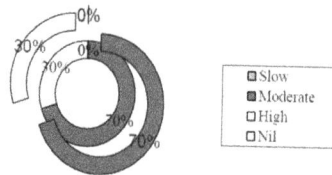

FIGURE 4.4
The growth rate in tea tourism. (Source: Self adaptation.)

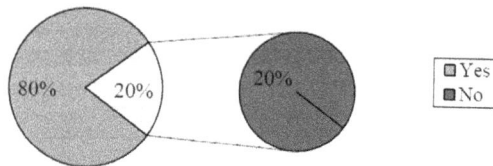

FIGURE 4.5
Tea tourism provides employment opportunities. (Source: Self adaptation.)

Tea tourism plays a significant role in promoting infrastructure development, women's empowerment, rural economic growth, and cultural exchange.

Third-world countries such as India benefit a great deal from tea tourism, and it does much good, such as opening up a local market for tea and providing seasonal employment. There are plenty of things you can do to expand your social contacts, such as making new friends, having countryside experiences, and participating in more cultural activities.

4.8 Potentially Problematic and Limiting Aspects of AI Use in the Travel Industry

Even as it has so much to offer in the travel industry, it has some challenges of its own that must be ironed out before it can realize its maximum potential. Some of those challenges in the use of AI in the travel industry are:

- Privacy issues arise as AI-driven systems tend to collect and process individuals' data in a bid to provide personalized recommendations. Because there is a risk of privacy problems, individuals do not wish to provide large amounts of personal data, such as their holiday interests (Wei et al., 2024). To maintain the trust of passengers, there is a need to strike the right balance between personalization and data privacy.
- Biased machines learned from unbalanced data can unwittingly perpetuate bias. This could lead to unequal or biased suggestions, which would affect the travel experience of certain groups and worsen economic disparities.
- While AI can be very useful, it may at times lack the human element of interacting with native experts or fellow visitors. When people in a society over-leverage technology usage, it may reduce the opportunity for meaningful encounters and people-to-people interaction.
- If tourists only heed AI suggestions, they might miss unique experiences and potential hidden sites because they are relying too much on technology.
- Smaller and less popular areas could be worse affected by the natural standardization of the visitor experience.
- AI systems only have the ability to suggest worthwhile options when they are fed with up-to-date and correct information. Incorrect or outdated information can misguide visitors and lead them to make inappropriate decisions.

- In terms of cultural sensitivity, AI-based systems cannot be fully aware of regional cultural nuances and sensitivities. The ultimate result could be suggestions that are contemptuous of or insensitive toward local traditions and cultures. Not everybody is comfortable with or used to AI-based platforms. Technological challenges may arise, as many older travelers or individuals from less technologically advanced regions may find the devices unclear or erratic.

4.9 Conclusion

Use of artificial intelligence (AI) for the creation of fully individualized itineraries according to each traveler's interests has the potential to transform the tea tourism business entirely by providing more relevant and up-to-date information about locations where tea is grown. The necessity of using AI in the tea tourism business is brought forward with these benefits and the success of using AI in many industries. Tourist sustainable development and economic empowerment of the citizens are two of the best advantages, well beyond travel advice. Raising awareness and solving the problems surrounding the use of AI in the tourism sector (Wei et al., 2024) is a priority. Problems of privacy, over-reliance on manufacturing, and algorithmic bias must be evaluated in depth. The use of technology has to not compromise human aspects of the experience's authenticity and focus. The balance between artificial intelligence capability and human intuition must be achieved if tea tourism is to utilize the potential of AI to the fullest extent. Firms, tourist boards, local governments, and artificial intelligence (AI) software firms need to work together in an attempt to cope with the dynamic tea tourism sector. What is required is the development of rules and ethics that will guarantee recommendations provided by AI are not only neutral and unbiased but also culturally tolerant. In order to make every visitor, technical expertise aside, enjoy the benefits of AI technology, proper investment must be made in digital literacy and infrastructure. In summary, AI could potentially transform the tea tourism sector, offering tourists an experience they will not soon forget and, in the process, boosting the economy of tea-producing regions. Being a specialty industry that encourages discovery, genuine relationships, and eco-sustainable behavior, the tea tourism business can benefit enormously from combining artificial intelligence with human understanding. Supposing that it has been working in partnership with AI for flexibility and ethical building, tea tourism is a motivating and innovative activity in our era.

References

Banerjee, S. & Tyagi, P. K. (2024). Implementation of digital technologies in promoting India as a tea destination. In Pankaj Kumar Tyagi, Vipin Nadda, Kannapat Kankaew, and Kaitano Dube (eds), *Dimensions of Regenerative Practices in Tourism and Hospitality* (pp. 259–268). IGI Global, Pennsylvania.

Bhatia, A. K. (2006). *Tourism Development, Principle and Practices*. Sterling Publisher Pvt, Ltd., New Delhi.

Bora, M. C. & Bora S. (2005). *The Story of Tourism, An Enhancing Journey through India's North East*. USB Publishers Distributions Pvt Ltd, New Delhi.

Das, A. K. (2009). Sustainability in tea industry: an Indian perspective. https://asr-jetsjournal.org/American_Scientific_Journal/article/view/1736

Das, H. N. (2008). *Assam Tea: Problems and Prospects. Origin and Development of Tea*. EBH Publisher, Guwahati.

Dutta, P., Kaushik, H., Bhuyan, R. P., Kaman, P. K., Kumari, A., Das, A., & Saikia, H. (2020). Relation of climatic parameter on tea production in organic condition specific to Assam. *International Journal of Current Microbiology and Applied Sciences*, 9(4), 2243–2249.

Goowalla, H. (2012). Labour relations practices in tea industry of Assam-with special reference to Jorhat district of Assam. *IOSR Journal of Humanities and Social Science*, 1(2), 35–41.

Goowalla, H., & Neog, D. (2011). Problem and prospect of tea tourism in Assam-a SWOT analysis. In *2011 International Conference on Advancements in Information Technology with Workshop of ICBMG*, University of Western Australia, Perth, Australia, 243–248.

Hazarika, K. (2011). Changing market scenario for Indian tea. *International Journal of Trade, Economics and Finance*, 2(4), 285.

Hazarika, M., & Muraleedharan, N. (2011). Tea in India: an overview. Two and a Bud, 58, 3–9.

Hussain, M. M. & Hazarika, S. D. (2010). Assam tea industry and its crisis. *Dialogue, (July-September)*, 12(1).

ITA (2009). Indian Tea Association, Tea Statistics, Surma Valley Branch, Silchar, Assam.

Jolliffe, L. (Ed.). (2007). *Tea and Tourism: Tourists, Traditions and Transformations*. Tourism and Cultural Change, Volume 11, Taylor and Francis, UK.

Jolliffe, L. (Ed.). (2010). *Coffee Culture, Destinations and Tourism*. Volume 24, Taylor and Francis.

Kim, K. H., & Park, D. B. (2013). Segmenting green tea consumers by purchase motivation in South Korea. *Journal of Agricultural & Food Information*, 14(2), 164–183.

Lee, J. (2007). *Tea and Tourism: Tourists, Traditions and Transformation*. Channel View Publications, Bristol.

Meshram, M., Singh, P., Sengupta, A., Expert, C. O. E. T. H., & Waknaghat, H. P. (2022). The development of digital technologies to upgrade the tea destination of Darjeeling and its effect on local communities–A critical analysis. *Development*, 30(3), 1613–1624.

Wei, Y., Wen, Y., Huang, X., Ma, P., Wang, L., Pan, Y., … & Wei, X. (2024). The dawn of intelligent technologies in tea industry. *Trends in Food Science & Technology*, 144, 104337.

5

Substation Level Short-Term Load Forecasting Using Graph-Based Signal Processing and Machine Learning Algorithms

Anantha Krishna Kamath, B. L. Rajalakshmi Samaga, and Lathika Jaganatha Shetty

5.1 Introduction to Load Forecasting

For energy suppliers and other players in the production, transmission, distribution, and markets for electric energy, electricity demand projections are crucial (Asbury, 1975; K. Srinivasan, 1975). A utility company's operations and planning depend on precise models for projecting electric power load. Load estimates are also crucial for energy suppliers and other players in the production, transmission, distribution, and markets for electric energy (Norma F. Hubele, 1990; Takeshi Haida, 1994).

Based on the duration for which demand is predicted, load forecasting is broadly classified into four types: Very Short-Term Load Forecasting, Short-Term Load Forecasting, Medium-Term Load Forecasting, and Long-Term Load Forecasting (Arunesh Kumar Singh, 2012). The period for Very Short-Term Load Forecasting (VSTLF) ranges from one minute up to one day (Luo, 2018). VSTLF is helpful for electric utilities and grid operators in making important decisions on real-time scheduling of electricity generation, real-time operations, demand response, security assessment, sensitivity analysis, and load frequency control (Luo, 2018). Short-Term Load Forecasting (STLF) plays an essential role in the operation of distribution substations. STLF provides hourly-based forecast results and spans from one day up to one week ahead (Rabindra Bahera, 2011; Arunesh Kumar Singh, 2012; Kong, 2017a,b). STLF is crucial for making decisions when the system is overloaded, for unit commitment, spinning reserve planning, and day-to-day system management. Medium-Term Load Forecasting (MTLF) covers a period from one month to one year, or up to 3 years (Xu, 2019; Arunesh Kumar Singh,

DOI: 10.1201/9781003506478-5

2012), whereas Long-Term Load Forecasting (LTLF) spans more than one year but less than 15 years (Arunesh Kumar Singh, 2012; Hong, 2019). MTLF and LTLF play important roles in maintenance scheduling, fuel reserve planning, unit commitment, energy contracts, load dispatching analysis, revenue from sales, load dispatching coordination, monthly peak load study, capacity expansion for electric utilities, network planning, capital investment, purchase of generating units, purchase of equipment, revenue analysis, and staff hiring. LTLF is a primary step taken by operators for planning the future generation, transmission, and distribution systems in power systems.

Data mining is the art and science of discovering knowledge, insights, and patterns in data. It involves extracting useful patterns from an organized collection of data.

The selection of load forecasting techniques and methods is based on the type of load forecasting. However, for all kinds of load forecasting, it is important to analyze historical data, which contains information such as date, time, demand, and more. It is also important to identify influencing factors such as temperature, season, and peak/off-peak times. These influencing factors depend on the geographical location of the substation for which demand is to be predicted. In this regard, an effort has been made to analyze and identify the influencing factors of the Talapady substation located in coastal Karnataka. The effectiveness of processed data is verified using different machine learning regressor models.

The major contributions of the chapter are as follows:

This chapter provides an idea of how to convert raw data into robust data for effective load forecasting. The conversion process, including the elimination of zero values and the transformation of outlier data into effective data, is explained in this work. This chapter focuses on processing raw data into an effective dataset and validating the processed data using different regressor models.

The rest of the chapter is organized as follows: Section 5.2 provides information on the processes involved in processing historical data, including zero elimination, data correction, and feature extraction. Section 5.3 focuses on the different regressor models used to validate the processed dataset. In Section 5.4, results after data processing and predicted load using regressor models are discussed. Finally, the conclusions of the proposed work are presented in Section 5.5.

5.2 Proposed Data Fusion

In this paper, electrical power consumption data of the Talapady distribution substation, located in Mangalore, from 2018 to 2022, are considered for data processing. Geographically, the substation is situated in a coastal area, between the Arabian Sea and the Western Ghats. The collected raw data consist of five features: year, month, date, time, and temperature, along with load. Hourly load consumption is recorded every day. From Figures 5.1 and 5.2, the minimum and maximum consumed loads are tabulated in Table 5.1. The distribution of raw data is not uniform, and there are some outlier values in the variation. Outlier values include zero or near-zero values on the lower side and values above 20 MW on the upper side.

For effective prediction using machine learning techniques such as multiple linear regression, decision tree regression, support vector regression, and random forest algorithms, outlier values must be either eliminated or modified, and the minimum number of features required is ten. In this chapter, outlier values are modified using the substitution method, and additional features are extracted using a graphical analysis method.

5.2.1 Feature Extraction

The process of turning unprocessed raw data into numerical features that can be processed while preserving the details of the original dataset is known as feature extraction. Predictions made with featured data are more accurate than those made using raw data with less detail. Raw data can be used to determine characteristics such as date, month, time of occurrence, and temperature. When more features are present in historical data, prediction methods using machine learning techniques like the Random Forest algorithm, decision tree algorithm, and support vector machine methods are more successful.

From the load dataset obtained from the substation, daily, weekly, monthly, and yearly averages of load were calculated to extract additional features from the data.

From Table 5.2, a steady growth in the yearly average is observed from 2018 to 2022. Consequently, the year has an impact on load forecasts at any given time. Thus, the year for which the load is predicted is considered one of the influencing features.

Geographically, the Talapady substation and its consumers are situated in the coastal part of Karnataka. In this region, the year is divided into two seasons: the rainy season from June to November and the non-rainy season from December to May. To consider the effect of season on load consumption, the monthly average load consumption is compared with the annual average load consumption. Referring to Table 5.3, in all the years, the monthly average from the sixth to the eleventh month is lower than the annual average, while the monthly average in the remaining months is higher than the annual average. Based on this comparison, an additional feature—season—can be considered, with season-1 representing the rainy season and season-2 representing the non-rainy season.

Table 5.4 provides a comparison between hourly load consumption and the daily average on 16/4/2022. According to the table and comparable

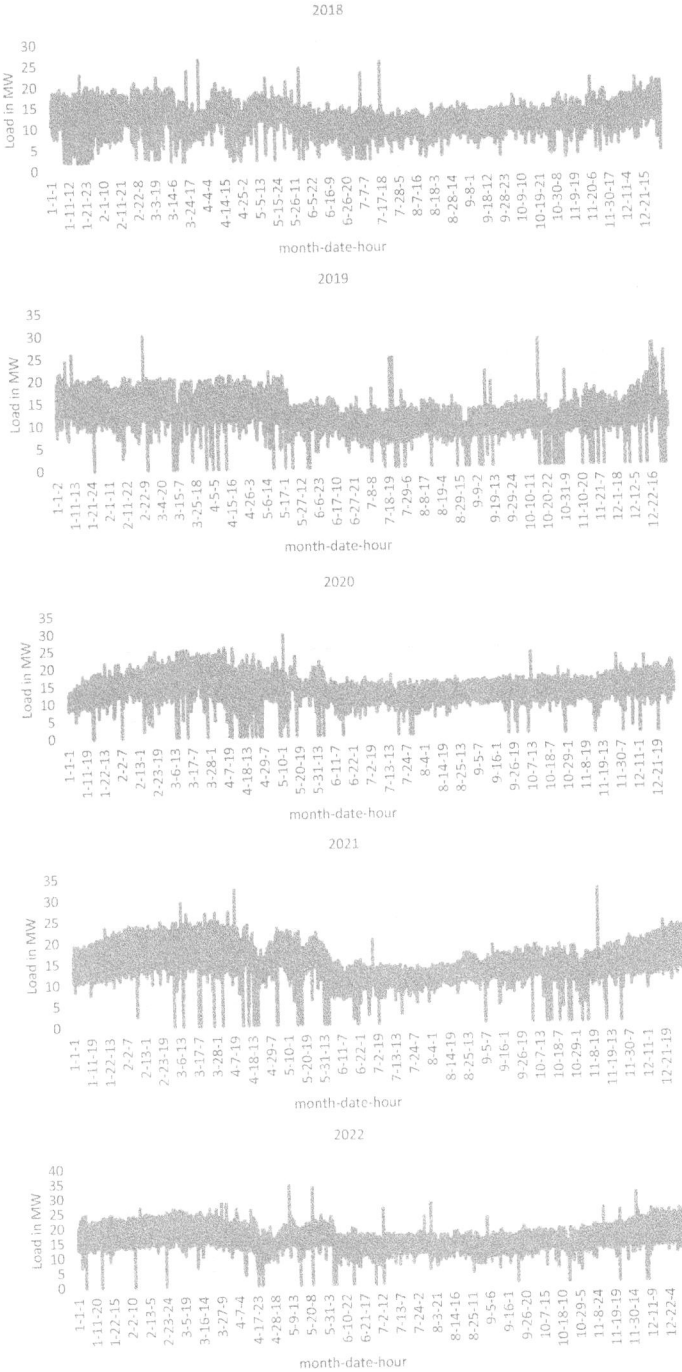

FIGURE 5.1
Hourly load variation of raw data for the years 2018, 2019, 2020, 2021, 2022.

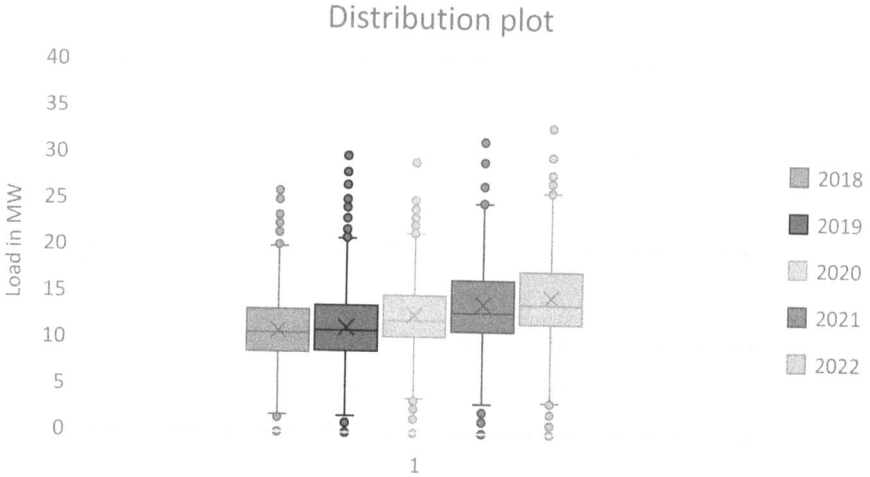

FIGURE 5.2
Distribution plot of raw load data for the years 2018, 2019, 2020, 2021, 2022.

TABLE 5.1

Year-Wise Minimum and Maximum Load Consumption

Year	Minimum Load in MW	Maximum Load in MW
2018	0	25.99
2019	0	29.82
2020	0	29.16
2021	0	32.14
2022	0	33.01

TABLE 5.2

Comparison of Yearly Average from 2018 to 2022

Year	Yearly Average
2018	10.91
2019	11.31
2020	12.50
2021	13.90
2022	14.73

observations made on other days, the load is significantly higher than the daily average from 7 a.m. to 10 p.m., while for the rest of the day it is lower than the daily average. Similar observations were found on other days as well. As a result, an additional feature based on load consumption at different time intervals—Peak and Off-Peak—can be identified.

TABLE 5.3

Comparison between Monthly Average and Yearly Average

Month	2018 Monthly Average	2018 Yearly Average	2019 Monthly Average	2019 Yearly Average	2020 Monthly Average	2020 Yearly Average	2021 Monthly Average	2021 Yearly Average	2022 Monthly Average	2022 Yearly Average
1	13.48		14.53		12.72		15.73		17.64	
2	14.22		14.19		15.11		17.77		18.89	
3	12.06		13.54		17.02		18.29		18.79	
4	12.10		14.15		15.07		15.80		14.56	
5	12.69		12.28		13.70		15.60		15.92	
6	8.99	10.91	9.54	11.31	11.31	12.5	10.55	13.90	12.36	14.73
7	8.76		8.11		10.50		10.18		11.91	
8	8.82		9.16		10.85		11.16		11.66	
9	9.34		8.42		11.15		11.86		11.95	
10	9.84		9.47		11.44		11.94		12.52	
11	10.62		10.88		11.52		12.41		13.91	
12	12.24		11.91		12.72		15.73		16.63	

Two features—day of the week and weekday/weekend—can be considered depending on the geography, the volume of residential, commercial, and industrial activity in the area, and the working culture of the region.

Every year, national festivals and some religious festivals occur on the same day. Thus, the day of the year can be considered as an additional feature. The proposed data fusion approach applied to raw data provides a robust and efficient dataset.

5.2.2 Proposed Method to Modify Outlier Values

Zero load or near-zero load recordings at the substation may occur due to power failures in the feeders caused by natural calamities. The duration of these power failures can be short or long, depending on their severity. To retain the originality of the dataset after modification, zero outlier values are modified based on the number of times zero load is recorded in a day. The conditions are

TABLE 5.4

Comparison between Hourly Variation of
Load and Daily Average on 16/4/2022

Hour	Consumed Load in MW	Daily Average in MW
1 am	12.94	
2	12.37	
3	12.3	
4	12.8	
5	13.14	
6	10.8	
7	17	
8	20.35	
9	22.13	
10	21.44	17
11	19.27	
12	17.94	
13	17.08	
14	17.95	
15	18.11	
16	18.87	
17	18.51	
18	18.22	
19	21.88	
20	21.64	
21	17.12	
22	17.69	
23	15.32	
24 midnight	13.17	

a. According to the recordings in Table 5.5, if the number of zero load recordings on a particular day is less than 25% of the total instances at which load consumption is recorded that day (i.e., a maximum of six zero recordings), then

$$L_{t0} = L_{da}$$

where L_{t0} is the zero load recording at time t on a particular day, and L_{da} is the daily average load calculated without considering the zero load recordings for that day. This adjustment is made to maintain the daily mean.

TABLE 5.5

Hourly Load Consumption on 16/7/2019

Year	Month	Date	Time in hours	Load in MW	Daily Mean
2019	7	16	1	3	
2019	7	16	2	4.41	
2019	7	16	3	4.38	
2019	7	16	4	4.34	
2019	7	16	5	4.21	
2019	7	16	6	3.89	
2019	7	16	7	4.77	
2019	7	16	8	4.71	
2019	7	16	9	4.82	
2019	7	16	10	2.44	
2019	7	16	11	0	
2019	7	16	12	0	6.69
2019	7	16	13	0	
2019	7	16	14	0	
2019	7	16	15	8.06	
2019	7	16	16	7.15	
2019	7	16	17	9.35	
2019	7	16	18	8.97	
2019	7	16	19	10.08	
2019	7	16	20	11.99	
2019	7	16	21	11.05	
2019	7	16	22	9.74	
2019	7	16	23	8.03	
2019	7	16	24	8.39	

b. On the other hand, if the number of zero load recordings on a particular day exceeds 25% of the total instances at which load consumption is recorded that day, as shown in Table 5.6, then

$$L_{t0} = L_{wa}$$

where L_{t0} is the zero-load recorded at time t on a particular day, and L_{wa} is the weekly average load for the week to which that day belongs, calculated without considering zero load recordings in that week. This adjustment is made to avoid deviation from the weekly average.

TABLE 5.6

Hourly Load Consumption on 7/3/2020

Year	Month	Date	Time in hours	Load in MW	Weekly Mean
2020	3	7	1	16.09	
2020	3	7	2	16	
2020	3	7	3	13	
2020	3	7	4	13	
2020	3	7	5	16	
2020	3	7	6	18.92	
2020	3	7	7	20.43	
2020	3	7	8	0	
2020	3	7	9	0	
2020	3	7	10	0	
2020	3	7	11	0	
2020	3	7	12	0	
2020	3	7	13	0	16.67
2020	3	7	14	0	
2020	3	7	15	0	
2020	3	7	16	0	
2020	3	7	17	0	
2020	3	7	18	0	
2020	3	7	19	0	
2020	3	7	20	21.65	
2020	3	7	21	21.45	
2020	3	7	22	20.89	
2020	3	7	23	20.02	
2020	3	7	24	18.87	

5.3 Proposed Load Forecasting Models

In this chapter, for effective load prediction, machine learning techniques including the multiple linear regression model, support vector regressor model, and random forest regressor model are proposed. The efficiency of the models is evaluated using the coefficient of determination.

5.3.1 Multiple Linear Regression Model

Multiple linear regression (MLR) is used to identify the relationship between one dependent variable and n independent variables. It determines the effect of each independent variable on the dependent variable individually and then predicts the dependent variable using the collective effect of all independent variables. In this paper, the consumed load is considered as the dependent variable (y), and ten parameters $x_1=$ year, $x_2=$ season, $x_3=$ month, $x_4=$ date, $x_5=$ week of the year, $x_6=$ weekday/weekend, $x_7=$ day of the week, $x_8=$ time, $x_9=$ peak/peak off time and $x_{10}=$ temperature—are considered as independent variables. The derived relationship for the dependent variable is

$$y = \beta_0 + \beta_1 x_1 + \beta_2 x_2 + \beta_3 x_3 + \beta_4 x_4 + \beta_5 x_5 + \beta_6 x_6 + \beta_7 x_7 + \beta_8 x_8 + \beta_9 x_9 + \beta_{10} x_{10}$$

β_0 is constant and β_i, $i=1$ to 10 is effect of x_i on y

5.3.2 Support Vector Regressor (SVR) model

Support vector regressor (SVR) is a type of support vector machine used for predictions. Support vector machines construct a hyperplane or a set of hyperplanes in a high-dimensional space to separate datasets. Kernel functions are used to construct these hyperplanes. There are different types of kernels in SVM, such as linear, polynomial, RBF (radial basis function), sigmoid, and precomputed. The regressor model finds a function that best predicts continuous values for a given set of inputs. In the proposed paper, the model is designed using the RBF kernel.

5.3.3 Random Forest Regressor (RFR) model

For a training dataset T with N tuples, the random forest regressor model works as below

- Create M bootstrap samples S_i ($1 \leq i \leq M$) of N tuples from the dataset using random sampling with replacement.
- For each sample, D decision trees are created. Each tree consists of a fixed number of features and a predefined depth. The tree recursively splits the dataset into two parts based on a certain criterion until a predefined stopping condition is reached.
- Based on the D decision trees, each sample predicts a value y_i

- The final prediction is generated by aggregating the predictions from all samples as $y_p = \dfrac{\sum\limits_{i=1}^{M} y_i}{M}$

  where M is the number of trees, and y_i is the predicted value from tree- i

In this chapter, the model is designed using random bootstrap samples, with 100 decision trees considered in each bootstrap sample.

5.3.4 Coefficient of Determination (R²)

The coefficient of determination is the proportion of variation in the dependent variable that can be predicted from the independent variables. It provides a measure of how well the dependent variable is replicated by the model.

$$R^2 = 1 - \frac{SS_{res}}{SS_{tot}}$$

where SS_{res}=sum of squares of residues and SS_{tot}=total sum of squares. It is more informative than MAE, MAPE, MSE, and RMSE

5.4 Results and Discussion

The dataset for the proposed work in this chapter is collected from the electrical power consumption data of the Talapady Substation, located in Mangaluru, from 2018 to 2022.

According to the raw data information shown in Table 5.7a, there are four parameters: date (in the format day-month-year), time, temperature, and load, with 73 pieces of information per day. After applying the proposed data fusion techniques to the raw data, the processed data, as referred to in Table 5.7b, contains 11 parameters: year, season, month, week of the year, date, weekday/weekend, day of the week, peak/off-peak load, time in hours, temperature, and daily load, with 264 pieces of information—approximately 3.5 times more than in the raw data. In Table 5.7b, under the weekday/weekend parameter, Monday to Friday is considered a weekday and represented as 1, while Saturday and Sunday are considered weekends and represented as 2. Under the day of the

TABLE 5.7A

Raw Data on 09/08/2022

Date	Time in hours	Temp	Load in MW
09/08/2022	1	26	10.18
	2	26	9.88
	3	26	10.03
	4	26	10.27
	5	26	10.57
	6	26	11.21
	7	26	11.92
	8	26	11.35
	9	27	9.63
	10	27	10.23
	11	27	8.66
	12	27	11.99
	13	27	12.72
	14	27	8.36
	15	28	10.3
	16	28	12.33
	17	28	7.89
	18	28	13.53
	19	28	15.47
	20	28	15.98
	21	27	13.21
	22	27	11.54
	23	27	9.06
	24	27	8.54

week parameter, Monday is represented as 1, Tuesday as 2, Wednesday as 3, Thursday as 4, Friday as 5, Saturday as 6, and Sunday as 7. In the peak/off-peak parameter, off-peak time is represented as 1, and peak time as 2.

Here, three different prediction models are designed using the processed dataset: multiple linear regression, support vector regression, and random forest regression. The designed models are tested using a sample dataset consisting of ten days of hourly load consumption data to provide a clear visualization of the differences between actual and predicted loads. The variations of actual and predicted demand by the different regressor models are shown in Figures 5.3–5.5.

TABLE 5.7B

Processed Data on 09/08/2022

Year	Season	Month	Week of the Year	Date	Weekday/Weekend	Day in a week	Peak/Off-peak	Time in hours	Temp	Load in MW
2022	2	8	32	9	1	4	1	1	26	10.18
2022	2	8	32	9	1	4	1	2	26	9.88
2022	2	8	32	9	1	4	1	3	26	10.03
2022	2	8	32	9	1	4	1	4	26	10.27
2022	2	8	32	9	1	4	1	5	26	10.57
2022	2	8	32	9	1	4	1	6	26	11.21
2022	2	8	32	9	1	4	2	7	26	11.92
2022	2	8	32	9	1	4	2	8	26	11.35
2022	2	8	32	9	1	4	2	9	27	9.63
2022	2	8	32	9	1	4	2	10	27	10.23
2022	2	8	32	9	1	4	2	11	27	8.66
2022	2	8	32	9	1	4	2	12	27	11.99
2022	2	8	32	9	1	4	2	13	27	12.72
2022	2	8	32	9	1	4	2	14	27	8.36
2022	2	8	32	9	1	4	2	15	28	10.3
2022	2	8	32	9	1	4	2	16	28	12.33
2022	2	8	32	9	1	4	2	17	28	7.89
2022	2	8	32	9	1	4	2	18	28	13.53
2022	2	8	32	9	1	4	2	19	28	15.47
2022	2	8	32	9	1	4	2	20	28	15.98
2022	2	8	32	9	1	4	2	21	27	13.21
2022	2	8	32	9	1	4	2	22	27	11.54
2022	2	8	32	9	1	4	1	23	27	9.06
2022	2	8	32	9	1	4	1	24	27	8.54

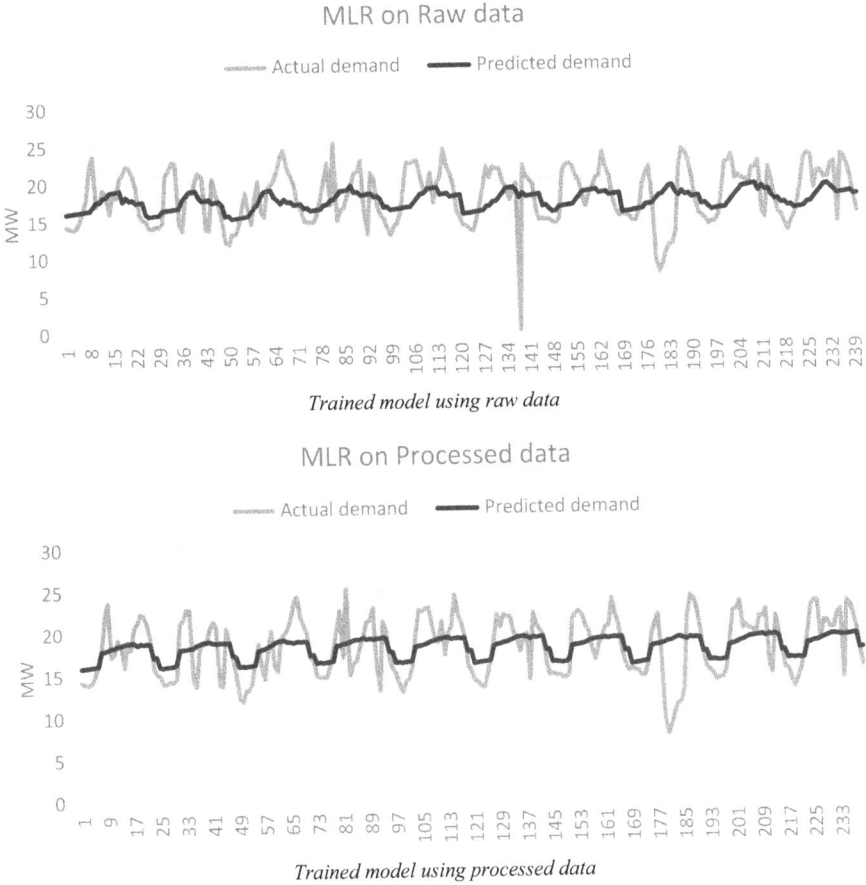

FIGURE 5.3
Actual and predicted demand by the multiple linear regressor model trained using raw data and processed data.

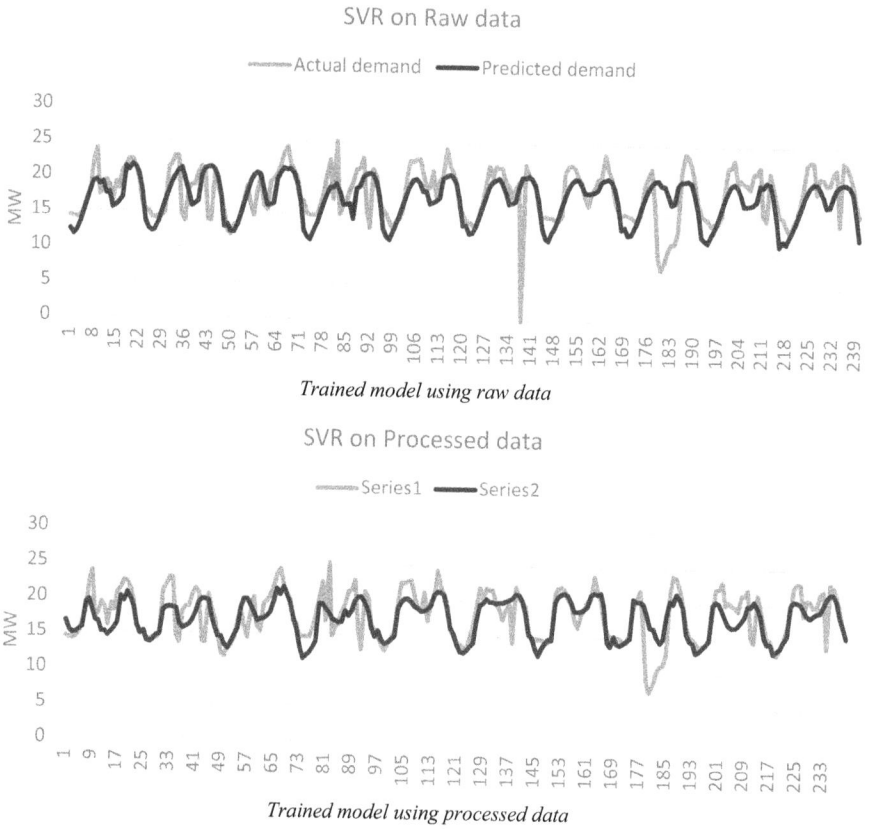

FIGURE 5.4

Actual and predicted demand by the support vector regressor model trained using raw data and processed data.

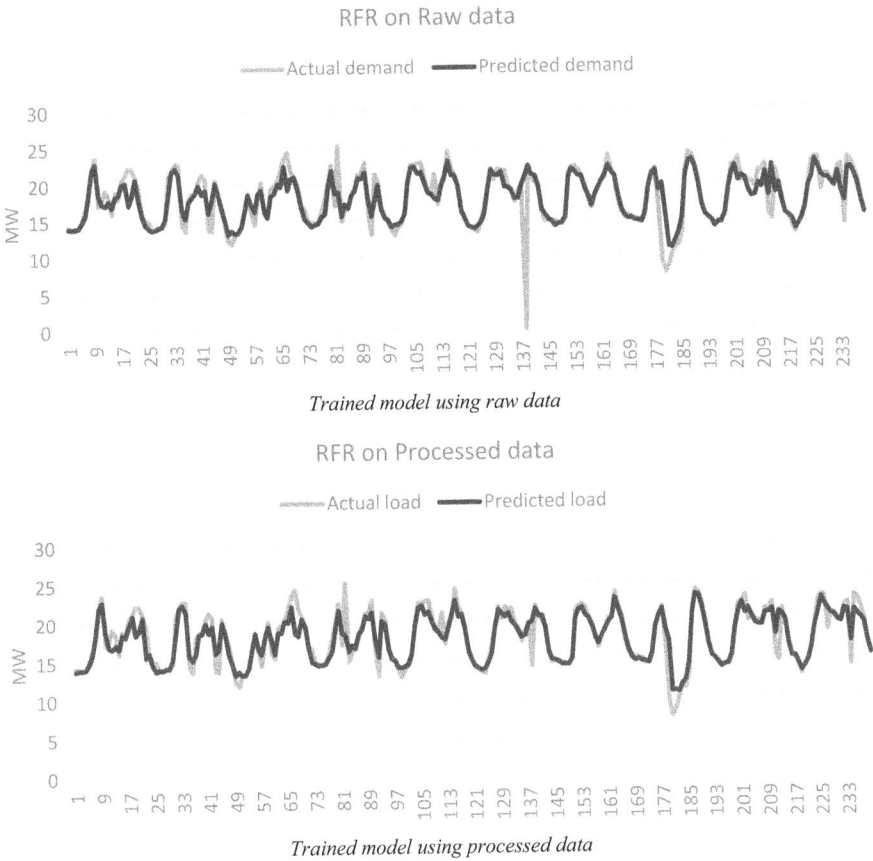

FIGURE 5.5

Actual and predicted demand by random forest regressor model trained using raw data and processed data.

TABLE 5.8

Comparison of R^2_score Obtained Using Different Models

	MLR Model		SVR Model		RFR Model	
	Training R^2_score	Testing R^2_score	Training R^2_score	Testing R^2_score	Training R^2_score	Testing R^2_score
Raw data	0.3409	0.3563	0.5858	0.6047	0.9632	0.749
Processed data	0.5355	0.5377	0.7402	0.7144	0.975	0.8213

However, to draw scientific conclusions from the study, an analysis of the model's accuracy is essential. In the proposed approach, the efficiency of the models is measured using the coefficient of determination (R^2_score) and R^2_score for different models, using both raw data and processed data, is tabulated in Table 5.8.

5.5 Conclusion

The proposed approach for short-term demand prediction is validated using a comprehensive dataset obtained from the Talapady substation, located in Mangaluru, Karnataka, India. The data fusion approach shows better results in terms of the coefficient of determination for all the proposed models in this paper. The results also indicate that the random forest regressor model exhibits a superior R^2_score compared to the multiple linear regression model and the support vector regressor model. Overall, the study concludes that the random forest regressor model applied to the processed dataset is superior to the other two models discussed in this chapter.

References

Arunesh Kumar Singh, I. S. (2012). Load forecasting techniques and methodologies: A review. *2nd International Conference on Power, Control and Embedded Systems* (pp. 1049–158). Allahabad, India: IEEE Explorer.

Asbury, C. (1975). Weather load model for electric demand and energy forecasting. *IEEE transaction on Power Apparatus and Systems*, 94, 1111–1116.

Hong, T. J. (2019). Global energy forecasting competition 2017: Hierarchical probabilistic load forecasting. *International Journal of Forecasting*, 35, 1389–1399.

K Srinivasan, R. P. (1975). Short term load forecasting using multiple correlation methods. *IEEE Transaction on Power Apparatus and Systems*, 1854–1858.

Kong, W. Z. (2017a). Short-term residential load forecasting based on LSTM recurrent neural network. *IEEE Transactions on Smart Grid*, 10, 841–851.

Kong, W. Z. (2017b). Short-term residential load forecasting based on resident behaviour learning. *IEEE Transactions on Power Systems*, 33, 1087–1088.

Luo, J. T. (2018). Real-time anomaly detection for very short-term load forecasting. *Journal of Modern Power Systems and Clean Energy*, 6, 235–243.

Norma F. Hubele, C.-S. C. (1990). Identification of seasonal short term load forecasting models using statistical decision functions. *IEEE transactions on Power Systems*, 5, 40–45.

Rabindra Bahera, B. P. (2011). A hybrid short term load forecasting model of an Indian grid. *Scientific Research, Energy and Power Engineering*, 3, 190–193.

Takeshi Haida, S. M. (1994). Regression based peak load forecasting using a transformation technique. *IEEE Transactions on Power Systems*, 9, 1788–1794.

Xu, L. S. (2019). Probabilistic load forecasting for buildings considering weather forecasting uncertainty and uncertain peak load. *Applied Energy*, 237, 180–195.

6

AI-Based Histopathological Image Analysis for Breast Cancer Diagnosis Using Deep and Machine Learning Algorithms

Rajwinder Singh, Hardeep Kaur, Jyoteesh Malhotra, Kuldeep Singh, and Gajendra Kumar

6.1 Introduction

Cancer of the breast (BC) is a leading cause of death worldwide in females, with a rising annual death rate (Balaji et al., 2023). The World Health Organization (WHO) reports that the prevalence of BC has sharply increased globally, accounting for 685,100 female fatalities and over two million new cases reported annually (Alirezazadeh et al., 2018). Breast cancer is the most frequent cancer among women, with more than seven million instances reported in the past 5 years alone (Aljuaid et al., 2022). There are two types of breast tumors (BTs): benign and cancerous. Malignant BTs can migrate to other regions of the body.

One important field of study for tissue data analysis is automated histopathological image categorization, which aims to enhance the process of decision-making for identifying diseases (Arooj et al., 2022). Pathologists can now identify diseases with increased accuracy thanks to the digitalization of tissue images, but automated tissue identification is still difficult because of the wide range of morphology and technical variances in photographs. Particularly in the recognition of brain tumors and the identification of Alzheimer's disease, DCNNs (Deep Convolutional Neural Networks) have greatly revolutionized radiology and medical research (Nisha et al., 2023). The development of computer-aided diagnostics (CAD) has been significantly impacted by these breakthroughs. Histopathological samples can be automatically classified into several BC subvariants using CAD programs, which are essential for medical studies and diagnostics. To reduce medical experts' efforts and enable precise diagnosis, this feature is crucial. The identification of breast cancer is still a difficult process despite advances in medical science because of an enormous number of

DOI: 10.1201/9781003506478-6

cases, protracted diagnostic phases, and challenges in deciphering biopsy data. By identifying BTs using EfficientNetB0 and five distinct pre-trained models—CNN, EfficientNetB1, ResNet50V2, MobileNetV3, and VGG16—this study seeks to overcome these issues. Several metrics are used to analyze the results, including ROC-AUC, precision, loss, and accuracy. Next, the outcomes are contrasted against results from related research using the BreakHis data.

The scientific literature presents a dynamic area of study on defining the types of BTs, demonstrating a range of approaches that effectively employ DL (deep learning) and visual evaluation to enhance diagnostic performance. Using seven distinct artificial intelligence algorithms on the BreakHis database, for instance, a study by Rana and Bhushan (2023) showed a substantial boost in accuracy for cancer classification, successfully processing imbalanced datasets without prior preprocessing. An improved STV (Swin-Transformer V2) was introduced by Kolla and Venugopal (2024) with the express purpose of classifying images of breast cancer histology into eight groups. This model uses multi-labeled statistics and incorporates a sigmoid activation mechanism to handle duplicates and incorrect data successfully. It also uses focal loss to make it more robust. Such groundbreaking studies have established a strong framework for further investigation, alongside those exploring quantum computing techniques and novel EfficientNetB0 architectures. Nonetheless, a thorough analysis of the literature highlights certain shortcomings in the application of innovative CNN algorithms and optimization methods for breast cancer (BC) categorization, particularly in recent studies. Despite considerable progress in research, there remains ample opportunity to enhance diagnostic efficiency and accuracy by integrating optimization techniques with contemporary machine learning algorithms.

To address these gaps and contribute innovatively to breast cancer (BC) treatment, our study leverages the latest advancements in EfficientNetB0 architectures and optimization techniques. This research aims to advance BC diagnostic techniques by comparing our method with both conventional and contemporary approaches. Additionally, it acts as a guide for upcoming research projects that attempt to improve the precision of healthcare diagnosis in the context of cancer therapy. The following is an overview of the presented paper's main contributions:

- Presenting a novel strategy that utilizes EfficientNetB0 and five other pre-trained models for diagnosing BTs.
- Impressive performance of EfficientNetB0: Surpassing previous state-of-the-art techniques, the EfficientNetB0 model achieved an exceptional accuracy rate of 99.50% in only ten iterations, representing a significant achievement.

- Various metrics, including accuracy, ROC-AUC curve, F1 score, and loss, were employed to comprehensively evaluate the models.
- A novel method for improving model resilience and preventing overfitting—a frequent problem in DL models—was demonstrated by using data enhancement techniques, such as stochastic brightness adjustments, flips, or rotations.

6.2 Literature Survey

Different research methodologies and procedures are covered in this section. We propose an advanced multiresolution feature visualization for eight classes, a novel approach for classifying BC using ANN and support vector machines. Umer et al. (2022) used an attribute combination and evaluation method with a 6B-Net complex CNN system to achieve multi-class cancer classification from pathology images. Their approach produced multi-class proficiency on histopathology slides of 90.00% on 8-class images and 94.20% on 4-class images. Murtaza et al. (2019) employed the Google Network architecture and utilized majority voting to classify histopathological images into subtypes. MUDeRN, using sections M and B, explored the application of ResNet to distinguish between benign and malignant breast tumor images and further classified them into their respective subtypes.

Joseph et al. (2022) created a BC diagnostic tool using the BreakHis database and data augmentation methods, such as rotations and shifts. With the use of Hu events, surfaces, and chromatic histograms, they physically identified characteristics and attained an impressive 97.87% accuracy at ×40 magnification. The results underlined the significance of customized model training by showing that learning at different resolution levels improves accuracy.

Using the BreakHis database with 400× magnification from an online source, a CNN-based approach was applied for early-stage BC identification. Among the configurations tested, the large networks achieved the highest accuracy of 90%, followed by NASNet-Large, DenseNet-201, and Inception-ResNetV3. Accurately identifying BC with the chosen ANN was the initial goal. Using transfer learning (TL) and a combination of deep neural networks (DNNs), including Inception-V3Net, ResNet-18, and ShuffleNet, Aljuaid et al. (2022) proposed a CAD method for BT categorization on the BreakHis database. The binary classification accuracies reported for ResNet, Inception-V3Net, and ShuffleNet were 99.7%, 97.66%, and 96.94%, respectively. For multiclass classification, the accuracy rates were 97.81%, 96.07%, and 95.79%.

Rana and Bhushan (2023) developed an automated tumor classification method that effectively managed unbalanced data using the BreakHis dataset without requiring preprocessing. Images were resized up to 224×224 dimensions and processed using seven different transfer learning models. According to the study, the Xception algorithm achieved the highest accuracy of 83.07%. The DarkNet53 model showed a notable result of 87.17% for best-balanced accuracy when tested with imbalanced data. For data with unequal classes, this research provides healthcare providers guidance on selecting appropriate models for cancer diagnosis. The AHT method was introduced by Iqbal et al. (2022) to improve CNN-based medical image categorization. On the BreakHis dataset, their method achieved 91.26% accuracy, and 93.21% on the scan dataset. Maan and Maan (2022) presented a deep learning (DL)-based saliency recognition method for breast cancer detection.

They utilized VGG16 and ResNet with the BreakHis database, attaining 96.7% accuracy during training and 90.4% accuracy during testing to identify and categorize cancerous areas across five diagnostic categories. Hirra et al. (2021) presented a breast cancer classification technique using histopathology images, combining DBN with LR. The system automatically extracted features from image patches, achieving an 86% accuracy rate in classification.

6.3 Methodology

This work presents an identification strategy that uses histopathology images from the BreakHis dataset and machine learning models to distinguish between cancerous and normal breast tissue (BC). As depicted in Figure 6.1, preprocessing—which includes resizing images, splitting data

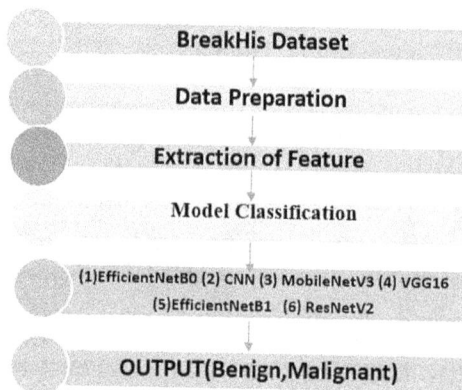

BreakHis Dataset

Data Preparation

Extraction of Feature

Model Classification

(1)EfficientNetB0 (2) CNN (3) MobileNetV3 (4) VGG16 (5)EfficientNetB1 (6) ResNetV2

OUTPUT(Benign,Malignant)

FIGURE 6.1
Flow chart for proposed algorithm.

into training and test sets, and balancing the dataset—is the primary step. Data enhancement approaches come next. The study employs a pre-trained model, EfficientNetB0, along with five other models—MobileNetV3, CNN, EfficientNetB1, ResNet50V2, and VGG16—for both feature extraction and classification. The effectiveness of these models is assessed using a variety of metrics, as detailed in the subsequent sections.

6.3.1 Acquiring Dataset

The information contained in BreakHis was obtained from an online source (Kaggle). It comprises 7,909 histological images of BT samples obtained from 82 patients. The distribution of images at each magnification level is detailed in Figure 6.2. These images have 700×460 pixel dimensions and are saved in PNG format with a 3-channel RGB color scheme. There are 2,480 normal and 5,429 malignant samples in the dataset, and each color channel consists of 8-bit resolution.

6.3.2 Data Preparation

One important element that impacts the effectiveness of algorithms using deep learning (DL) is data pretreatment. To make the system's interpretation easier, it entails resizing images, balancing the dataset, and organizing the data. The following techniques for preprocessing are used in the current method:

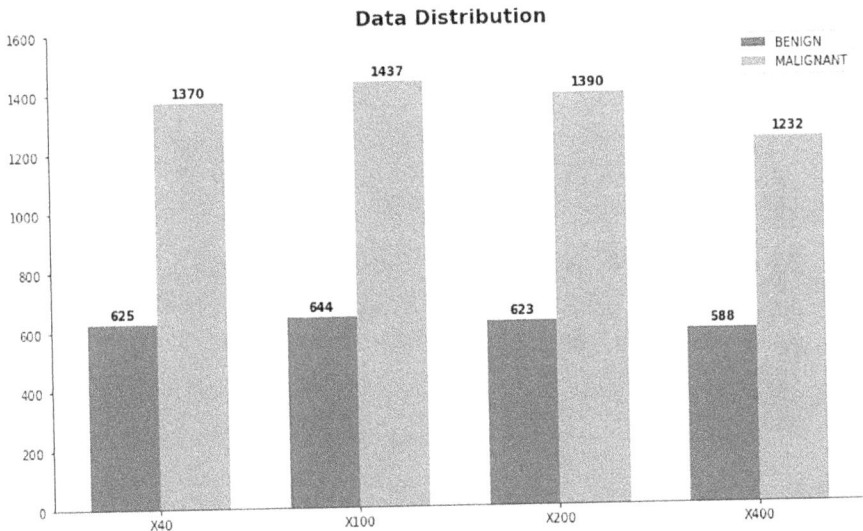

FIGURE 6.2
Data distribution.

- **Train-Test Separation**

  EDA, or exploratory data analysis, is a crucial stage in the data analysis process that entails identifying and comprehending a dataset's underlying patterns, correlations, and properties. Throughout the study, EDA aids decision-making by offering insightful analyses of the data. Five folds, one for training and one for testing, make up the data collection used in this investigation. The third fold, which has the highest percentage of training images, was used in the suggested technique to guarantee that training and testing data originate from different individuals. As shown in Figure 6.3, this method decreased the possibility of overfitting and provided a precise evaluation of the model's capacity for generalization.

- **Train Set Balance**

  There is a glaring imbalance in the groups within the chosen second fold: the number of tumors (3,630) greatly exceeds the number of benign instances (1,702). Inaccurate categorization may result from these differences, as shown in Figure 6.4. Using an upsampling approach, the benign and cancerous classes are balanced. Specifically, scaling is applied to equalize the class distribution by increasing the number of instances in the benign group (the minority class). This approach is employed to retain potentially valuable information from the larger class (the cancerous class) without discarding it.

- **Enhancement of Data**

  An intended application of data augmentation can improve classification accuracy, increase the total number of samples, and counteract overfitting. In addition to the basic augmentation techniques, color adjustments such as luminosity changes, flipping, and rotation

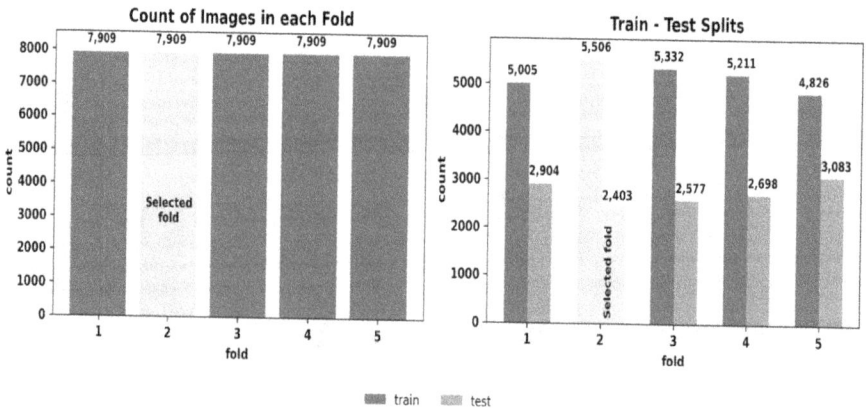

FIGURE 6.3
Train-test splitting of the dataset.

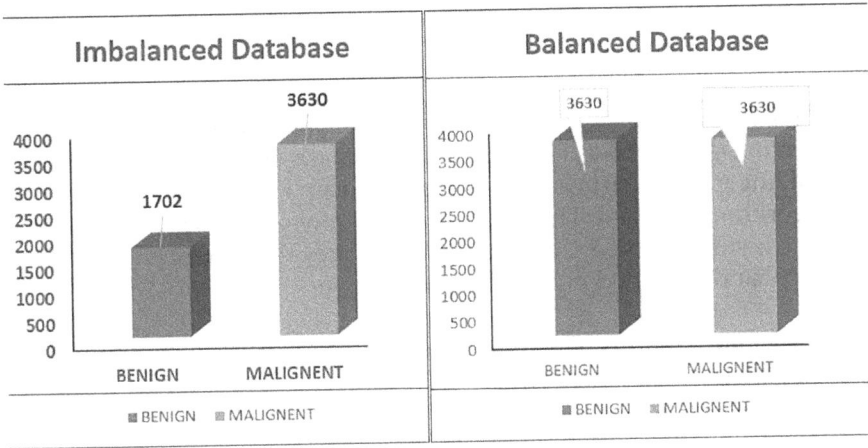

FIGURE 6.4
Balancing dataset.

are used to enhance the data. Randomly rotating an image can result in zero-pixel borders around the corners. By addressing overfitting in DCNN models, image augmentation contributes to data growth.

6.3.3 Feature Extraction

EfficientNetB0 uses methods such as depth-wise transformations to efficiently extract features while using minimal computational resources. The VGG16 algorithm employs convolutional neural networks in conjunction with max pooling and ReLU activation to gather and organize features. Finally, ResNet50V2 uses the initial layers and "remaining blocks" for pattern refinement to accelerate feature extraction and ensure continuous data flow.

6.4 Results

$$\text{Accuracy} = \frac{TP + TN}{TP + FP + FN + TN} \qquad (6.1)$$

$$\text{Precision} = \frac{TP}{TP + FP} \qquad (6.2)$$

$$\text{F1} - \text{Score} = 2 \times \frac{TP}{2} \times TP + FP + FN \qquad (6.3)$$

A visual aid for evaluating a classification model's accuracy across a range of threshold values is the ROC-AUC. The True Positive versus False Positive Rate plot provides information on how well the model performs under various classification settings. This helps in comprehending the trade-off between mistakenly classifying negative instances as positive and correctly recognizing positive instances. In the field of deep learning, various types of loss algorithms are used. The binary cross-entropy loss is widely applied for binary classification tasks. This metric measures the discrepancy between the actual and predicted probability distributions.

- **Efficient Net**: EfficientNet utilizes an integrated scaling approach to provide modern accuracy with greater efficiency while balancing dimensions, size, and sharpness. Its design aims to achieve higher computational efficiency than previous versions.
- **VGG16**: The 16-layer deep neural network model is well known for its reliable effectiveness in image classification tasks, even though it has many parameters and demands a lot of processing resources. It is also user-friendly.
- CNNs are neural network models particularly useful for processing visual data. The convolutional layers of CNNs are utilized to recognize features. Applications such as video and image recognition use pooling layers to reduce the number of pixels.
- **Mobile Net**: By utilizing depth-wise separable convolutions to reduce computational costs and parameters, the MobileNet family of lightweight CNNs is designed for efficient operation on mobile and embedded systems.
- **ResNet50**: The ResNet50 deep neural network, which has demonstrated impressive results in image classification applications, is a 50-layer CNN that employs skip connections and residual learning to address the vanishing gradient problem.

6.4.1 Oservations and Discussions

Despite this, EfficienNetB0 performs better than the other five pre-trained models. With a score of 98%, Figure 6.5 demonstrates that EfficientNetB0 surpasses the other models in terms of ROC-AUC. In contrast, the ROC-AUC values for CNN, VGG16, EfficientNetB1, ResNet50V2, and MobileNetV3 were 89%, 84%, 85%, 83%, and 74%, respectively. These results indicate that EfficientNetB0 outperforms the other models in this study.

Figure 6.6 provides a comparative analysis of the accuracy results for each of the six approaches used in the primary method. While the other models achieved an accuracy rate of 78%, the EfficientNetB0 model stands out with exceptional performance, achieving 99.50%.

Test ROC-AUC

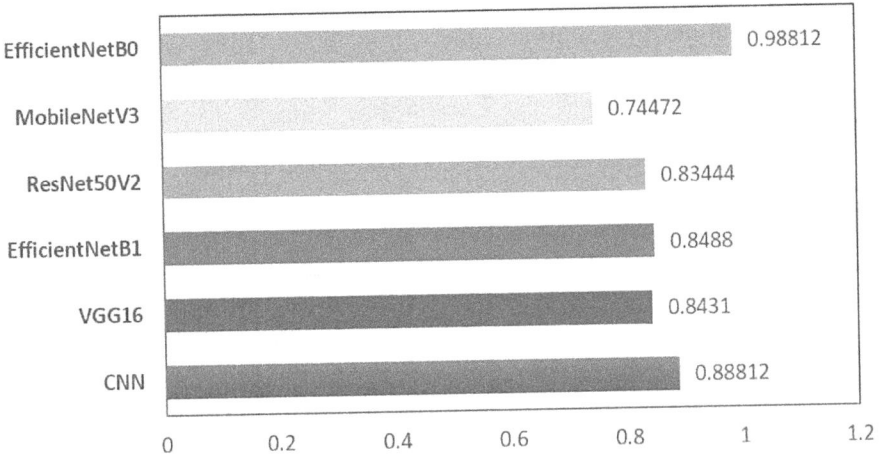

EfficientNetB0	0.98812
MobileNetV3	0.74472
ResNet50V2	0.83444
EfficientNetB1	0.8488
VGG16	0.8431
CNN	0.88812

FIGURE 6.5
The test ROC-AUC results of the DL models for breast cancer detection using the BreakHis dataset.

Test Accuracy

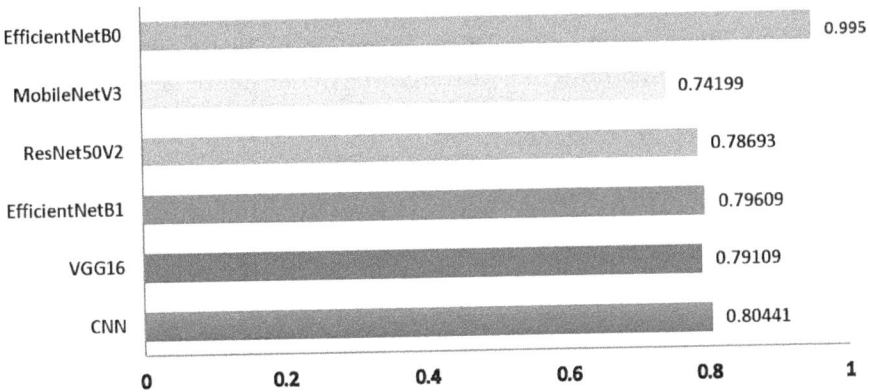

EfficientNetB0	0.995
MobileNetV3	0.74199
ResNet50V2	0.78693
EfficientNetB1	0.79609
VGG16	0.79109
CNN	0.80441

FIGURE 6.6
The Accuracy of the different models for breast cancer detection using the BreakHis dataset.

Figure 6.7 presents a comparison of the loss outcomes for the six models in the main technique. While the other models achieved a loss of at least 42%, the EfficientNetB0 model achieved a remarkably low loss of 0.0171%, demonstrating outstanding performance.

In EfficientNetB0, using 10 epochs, the model accurately identified 287 cases of cancer (TP) and 127 normal instances (TN). However, there were two

Test Loss

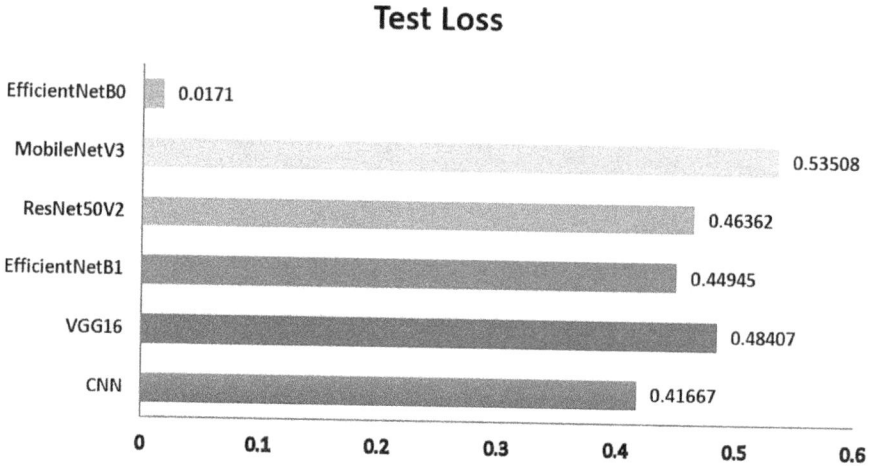

Model	Test Loss
EfficientNetB0	0.0171
MobileNetV3	0.53508
ResNet50V2	0.46362
EfficientNetB1	0.44945
VGG16	0.48407
CNN	0.41667

FIGURE 6.7
Loss bar graphs for different models.

cancerous samples incorrectly identified as benign (FN), and benign samples were mistakenly classified as malignant (FP), as shown in Figure 6.8. Overall, EfficientNetB0 demonstrates superior performance in terms of correct classifications compared to the other models.

Figure 6.9 illustrates 369 normal samples erroneously identified as cancerous and 251 carcinoma samples mistakenly classified as normal. Using MobileNetV3, the algorithm correctly predicted 355 instances as normal (TN) and 1,428 instances as cancerous (TP).

6.5 Final Reflections and Future Directions

This study introduces a proposed approach for breast cancer (BC) classification using EfficientNetB0 alongside five other pre-trained models. The EfficientNetB0 model notably achieved an accuracy of 99.50% within ten iterations, as shown in Table 6.1, marking a promising initial outcome. Research conducted on the BreakHis dataset demonstrates that this result not only surpasses traditional approaches but also highlights the effectiveness and importance of these optimized strategies in advancing tools for BC. Despite these promising results, there are several key considerations for future research. Firstly, we have not yet assessed our model's effectiveness in actual BT categorization situations, which is a critical step that we are currently omitting. Moreover, the training and optimization of our system demand substantial computational resources, which may be challenging to access in low-resource areas, potentially hindering the widespread application of our

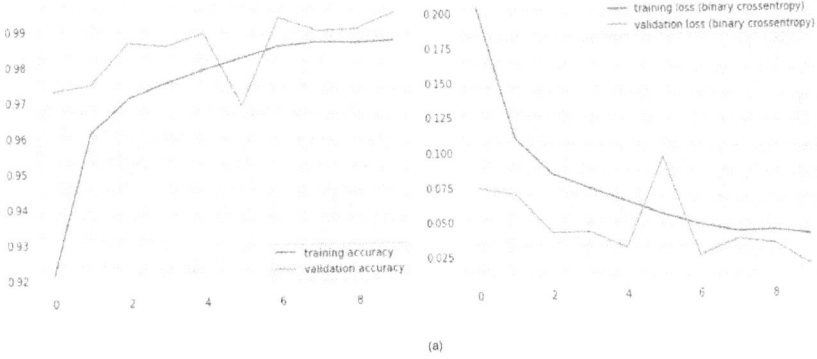

(a)

Confusion Matrix

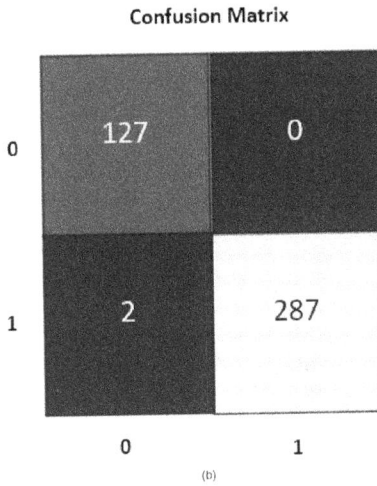

(b)

FIGURE 6.8
Confusion matrix and training (accuracy and loss) curve for best model (EfficientNetB0).

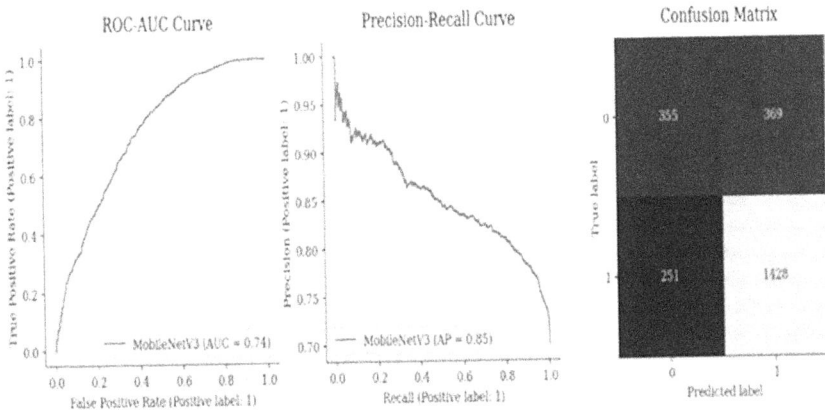

FIGURE 6.9
Confusion matrix for the mobileNetV3 model.

TABLE 6.1

Comparison of the Proposed Model with Previous Research Work on the BreaKHis Dataset

Author's	Dataset Used	Classifier	Performance
Alirezazadeh et al. (2018)	BreakHis	MobileNet	Accuracy = 88.50
Kassani et al. (2019)	BreakHis	VGG19	Accuracy = 99%
Sharma and Mehra (2020)	BreakHis	Layer-wise fine-tuning	Accuracy = 89.31%
Umer et al. (2022)	BreakHis	6B-Net	Accuracy = 94.20%
Maleki et al. (2023)	BreakHis	Alex Net network	Accuracy = 91.92%
Jacinta and Devi (2024)	BreakHis	Dense Net 121	Accuracy = 96.09%
Proposed model	BreakHis	EfficientNetB0	Accuracy = 99.50%

approach. Additionally, although our model outperforms others, it is essential to contextualize these achievements within the broader landscape of diagnostic tools.

References

Alirezazadeh, P., Hejrati, B., Monsef-Esfahani, A., Fathi, A. (2018). Representation learning-based unsupervised domain adaptation for classification of breast cancer histopathology images. *Biocybern. Biomed. Eng.* 38(3), 671–683. https://doi.org/10.1016/j.bbe.2018.04.008

Aljuaid, H., Alturki, N., Alsubaie, N., Cavallaro, L., Liotta, A. (2022). A. Computer-aided diagnosis for breast cancer classification using deep neural networks and transfer learning. *Comput. Methods Progr. Biomed.* 223, 106951. https://doi.org/10.1016/j.cmpb.2022.106951

Arooj, S., Atta-Ur-Rahman, Zubair, M., Khan, M. F., Alissa, K., Khan, M. A., Mosavi, A. (2022). A Breast cancer detection and classification empowered with transfer learning. *Front. Public Health* 10, 1. https://doi.org/10.3389/fpubh.2022.924432

Balaji, S., Arunprasath, T., Rajasekaran, M. P., Vishnuvarthanan, G., Sindhuja, K. (2023). Computer-aided diagnostic system for breast cancer detection based on optimized segmentation scheme and supervised algorithm. *Automatika* 64, 1244–1254. https://doi.org/10.1080/00051144.2023.2244307

Hirra, I., Ahmad, M., Hussain, A., Ashraf, M. U., Saeed, I., Qadri, S. F. (2021). Breast cancer classification from histopathological images using patch-based deep learning modeling. *IEEE Access* 9, 24273–24287. https://doi.org/10.1109/ACCESS.2021.3056516

Iqbal, S., Qureshi, A. N., Ullah, A., Li, J., Mahmood, T. (2022). Improving the robustness and quality of biomedical cnn models through adaptive hyperparameter tuning. *Appl. Sci.* 12, 11870. https://doi.org/10.3390/app122211870

Jacinta, P., Devi, S. S. (2024). Classification of breast cancer histopathological images using transfer learning with DenseNet121. *Procedia Comput. Sci.* 235, 1990–1997. ISSN 1877-0509. https://doi.org/10.1016/j.procs.2024.04.188

Joseph, A. A., Abdullahi, M., Junaidu, S. B., Ibrahim, H. H., Chiroma, H. (2022). Improved multi-classification of breast cancer histopathological images using handcrafted features and deep neural network (dense layer). *Intell. Syst. Appl.* 14, 200066. https://doi.org/10.1016/j.bspc.2023.105152

Kassani, S. H., Kassani, P. H., Wesolowski, M. J., Schneider, K. A., Deters, R. (2019). Classification of histopathological biopsy images using ensemble of deep learning networks. arXiv preprint, arXiv:1909.11870. https://doi.org/10.48550/arXiv.1909.11870

Kolla, B., Venugopal, P. (2024). An integrated approach for magnification independent breast cancer classification. *Biomed. Signal Process. Control* 88, 105594. https://doi.org/10.1016/j.bspc.2023.105594

Maan, J., Maan, H. (2022). Breast cancer detection using histopathological images. Preprint at https://arxiv.org/abs/2202.06109

Maleki, A., Raahemi, M., Nasiri, H. (2023). Breast cancer diagnosis from histopathology images using deep neural network and xgboost. *Biomed. Signal Process. Control* 86, 105152. https://doi.org/10.1016/j.bspc.2023.105152

Murtaza, G., Shuib, L., Wahab, A. W. A., Mujtaba, G., Nweke, H. F., Al-Garadi, M. A., Zulfiqar, F., Raza, G., Azmi, N. A. (2019). Deep learning-based breast cancer classification through medical imaging modalities: State of the art and research challenges. *Artif. Intell. Rev.* 53, 1655–1720. https://doi.org/10.1007/s10462-019-09716-5

Nisha, A. V., Rajasekaran, M. P., Kottaimalai, R., Vishnuvarthanan, G., Arunprasath, T., Muneeswaran, V. (2023). Hybrid d-ocapnet: Automated multi-class Alzheimer's disease classifcation in brain MRI using hybrid dense optimal capsule network. *Int. J. Pattern Recognit. Artif. Intell.* 37, 2356025. https://doi.org/10.1142/S0218001423560256

Rana, M., Bhushan, M. (2023). Classifying breast cancer using transfer learning models based on histopathological images. *Neural Comput. Appl.* 35, 14243–14257. https://doi.org/10.1007/s00521-023-08484-2

Sharma, S., Mehra, R. (2020). Efect of layer-wise fne-tuning in magnifcation-dependent classifcation of breast cancer histopathological image. *Vis. Comput.* 36, 1755–1769. https://doi.org/10.1007/s00371-019-01768-6

Umer, M. J., Sharif, M., Kadry, S., Alharbi, A. (2022). A. Multi-class classifcation of breast cancer using 6b-net with deep feature fusion and selection method. *J. Person. Med.* 12, 683. https://doi.org/10.3390/math10214109

7

AI-Optimized Active Cell Balancing Technique Using Single Inductor for Intelligent Battery Management in IoT-Enabled Energy Systems

Aman Sharma, Shelly Vadhera, and Sunitha Cheriyan

7.1 Introduction

Individual cells are arranged in parallel and series to form a Li-ion battery. The battery cell with the lowest capacity, which is most prone to overcharging or overdischarging, determines the amount of energy that can be charged and discharged from the battery pack. Cells with larger capacities may only undergo partial cycles. This is because each battery cell has a unique capacity and may be in different states of charge. For higher-capacity cells to complete their full charge and discharge cycles without overcharging or overdischarging other cells in the pack, battery balancing is required. Voltage equalization, also known as SOC equalization, is one of the two methods to achieve battery balancing (Vardhan et al., 2017). According to this research, energy can be transferred from a high-SOC cell to a low-SOC cell until the SOC of every cell equals the average SOC of the battery pack. This is referred to as SOC-based cell balancing.

There are two main methods for balancing SOC: passive cell balancing and active cell balancing. Passive cell balancing uses resistors and a combination of switches or diodes to balance the SOC of the cells. This approach dissipates excess energy as heat through the resistor. While cost-effective, it is inefficient due to energy loss as heat, and it is only effective during charging, with limited or no use during discharging. In contrast, active cell balancing uses components like inductors and capacitors to transfer charge from one cell to another and balance the entire battery pack. Since one cell's energy is used to charge another, this process is more efficient.

DOI: 10.1201/9781003506478-7

This study proposes a method to balance n cells using $2n$ switches and diode combinations with a single inductor. To balance the SOC of each cell with that of the battery pack, switches will be turned on and off based on measurements and comparisons of the SOC of each cell.

7.2 Theory

Cell balancing plays a key role in Battery Management Systems (BMS). It reduces safety hazards and increases the longevity and performance of the system. Different methods, known as "passive cell balancing" and "active cell balancing," are used to balance the state of charge (SoC) of a cell. In passive cell balancing, resistance is employed as a means of achieving cell equilibrium. Overcharged cells are balanced, and excess energy is dissipated through resistance (Mohamed Dawod et al., 2011). However, this process wastes energy, making it inefficient. Excessive heat generation can also cause issues for the system. Active cell balancing is an alternative technique for achieving cell equilibrium. One method used in active cell balancing involves capacitors and inductors to transfer energy from one cell to another. Research has been conducted on the effectiveness of capacitor-based active cell balancing techniques. However, there are several drawbacks to this approach, including transient events during the charging and discharging phases and considerable energy loss (Marcin et al., 2024). An inductor can be used as an additional means of balancing the cell. This battery system contains N cells, and $2N+2$ switches are utilized. To prevent short circuits, each switch is connected to an antiparallel diode.

7.2.1 Passive Cell Balancing

Passive cell balancing is a well-liked option for numerous battery applications due to its affordability and ease. This method is more economical compared to other techniques. Resistance of different ratings is used according to the needs of the system. Their circuitry is simple and easy to understand.

Consumer devices, including cellphones, computers, tablets, and portable power banks, frequently use passive cell balancing. Most of these gadgets run on lithium-ion batteries, which need to be managed well to guarantee longevity and safety. Passive balancing is a cost-effective and simple approach that is well-suited to the small size and low capacity of these battery packs.

7.2.1.1 Principle of Passive Cell Balancing

Passive cell balancing works by dissipating excess energy through resistance, known as bleed resistance, in the form of heat. When a cell reaches an SOC greater than that of other cells in the battery pack (Mohamed Dawod et al., 2011), the resistance connected in parallel with that cell draws current until the SOC of that cell reaches the same value as that of the other cells. This method is straightforward and easy to implement. Passive balancing ensures uniform voltage levels across all cells, enhancing battery pack performance and longevity.

7.2.2 Active Cell Balancing

Active cell balancing is a sophisticated method used to equalize the charge among cells in a battery pack by redistributing energy from higher-charged cells to lower-charged ones. By using this method instead of passive ones, energy losses are minimized, and balancing efficiency and efficacy are increased (Duraisamy & Kaliyaperumal, 2020; Shah et al., 2018).

7.3 Literature Review

The past study highlights the importance of cell balancing in enhancing battery performance and longevity. While passive balancing methods offer simplicity and cost benefits, active balancing methods are superior in terms of efficiency and speed. The choice of a balancing technique depends on specific application requirements, including cost constraints, energy efficiency, and system complexity. Future research is directed towards optimizing these balancing techniques, improving control algorithms, and integrating advanced materials to further enhance the performance of battery management systems (Daowd et al., 2011). Another paper proposes a simple, high-performance battery balancing mechanism for a series-connected Li-ion battery. Compared with the conventional battery balancing method, the proposed circuit and controller are more straightforward, easier to use, and robust. Here, proportional derivative (PD) control and a fuzzy logic controller (FLC) control the circuit balancer to ensure all series batteries are equalized (Safitri et al., 2022). Furthermore, another investigation mentioned that the key component of the battery management system (BMS), which is used to increase battery run duration and service life, is cell balancing. The need for larger and higher-performing battery packs is driving attention toward various cell balancing five approaches. The most widely used method is passive balancing due to its ease of use and inexpensive cost. In order to maintain the BMS board temperature within a reasonable range, the balancing system must have an adequate thermal scheme because the balancing energy is released as heat by the balancing resistors. This study proposes a machine learning (ML)-based balancing control algorithm to pick the balancing resistor optimally in terms of temperature rise, balancing time, C-rate, and degree of cell imbalance. The ML-based balancing control algorithm is proposed to improve the balancing time and optimize power loss management. Variable resistors are utilized in the passive balancing system in order to optimize the power loss and obtain optimal thermal characterization (Duraisamy & Kaliyaperumal, 2021). In another study, a double-tiered cell equalization system is designed by taking into account the effects of six parasitic resistances of inductors and capacitors used in buck-boost and switched-capacitor kinds of cell balancing systems. This system's first layer is a buck-boost cell equalizer, and its second tier—which has a fast-equalizing speed—is implemented by a switched-capacitor equalizer. One pair of complementary square wave signals controls a common switch diagram shared by the two layers. As a result, there are far fewer switches needed for complicated closed-loop control, and neither voltage nor current signals need to be monitored. Detailed illustrations are provided for circuit configuration, modeling, and design concerns. The suggested cell balancing system's viability is shown by the results of both simulations and experiments.

7.4 Novelty

In this work, the single-inductor balancing technique is applied with $2n$ switches and diodes. Additionally, in contrast to other approaches, the number of switches utilized in the procedure is lowered to three in the event that adjacent cells need to be balanced. In contrast to the above methods, which utilize $2n+2$ switches to balance the same number of cells, this strategy uses fewer switches.

7.5 Purposed Circuit

As seen in Figure 7.1, the proposed single-inductor circuit will use a single inductor to charge one cell from another. Using an SOC-based function controller, the SOC of each cell will be measured in this manner, and the balancing circuit's switches will be turned ON or OFF correspondingly. First, an overcharged cell will send its energy to an inductor, which will then use a collection of switches to transfer energy to a less charged cell, as seen in the flow diagram that follows (Figure 7.2).

Assuming that, as shown in Figure 7.3, Cell 1 has more energy than Cell 2. Cell 1 will charge the inductor with polarity as positive on the upper side, as shown in Figure 7.4 after Time Period t', at which point the inductor polarity will be reversed, and switches S3 and S4 will be turned ON. Because the inductor doesn't allow a sudden change of current, and current remains to

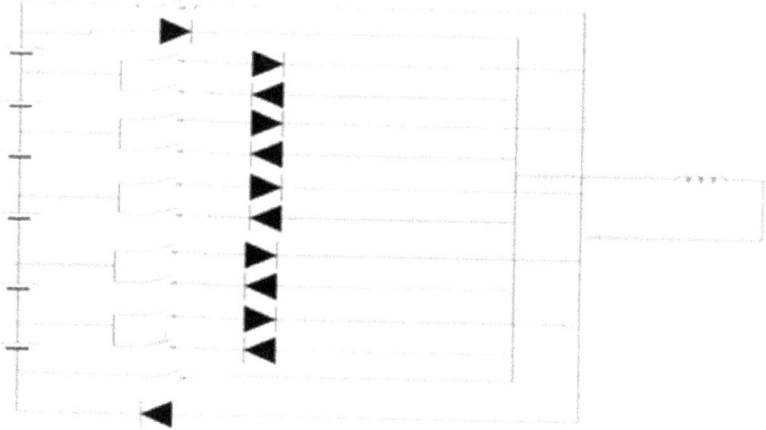

FIGURE 7.1
Cell balancing technique using inductor.

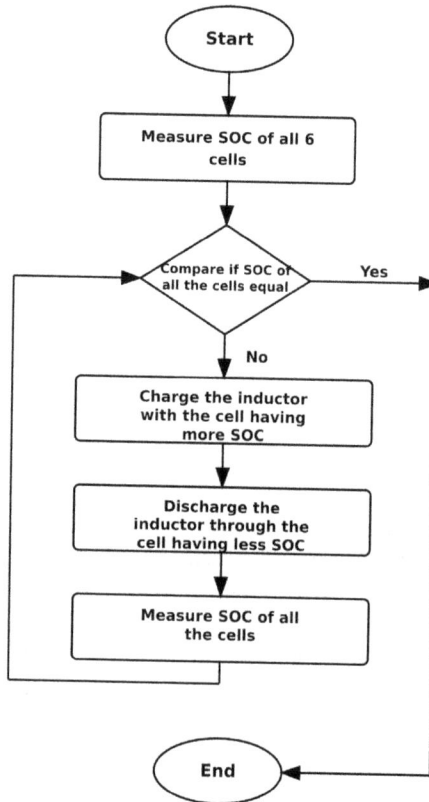

FIGURE 7.2
Flowchart for a function block.

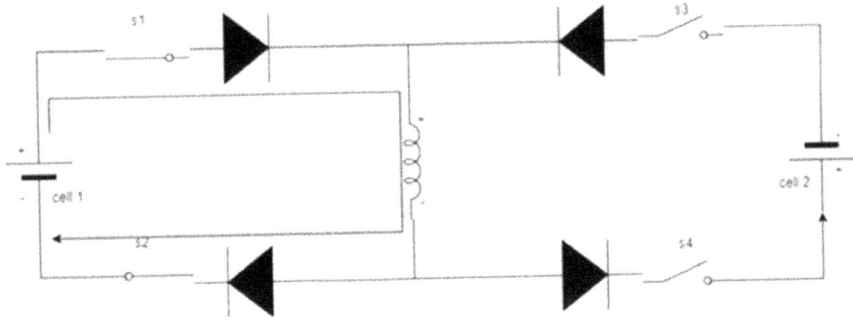

FIGURE 7.3
Current direction from cell to inductor.

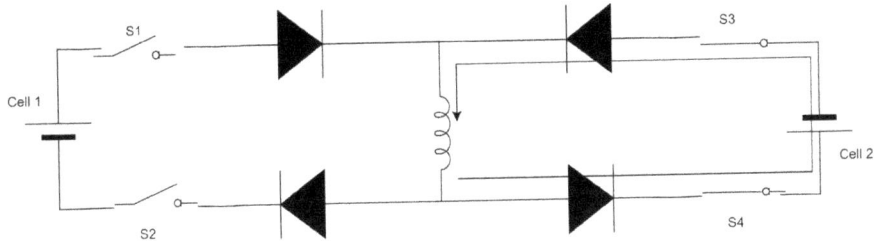

FIGURE 7.4
Current direction from inductor to cell.

flow in the same direction, and it will continuously charge Cell 2 gradually during Time Period t''. Thus, T is the entire amount of time that the circuit needs to transfer energy between cells.

$$T = t' + t''$$

7.6 MATLAB® Circuit

The MATLAB connection will be made as indicated in Figure 7.5. The function block algorithm in MATLAB, depicted in the flow diagram in Figure 7.2, will be used.

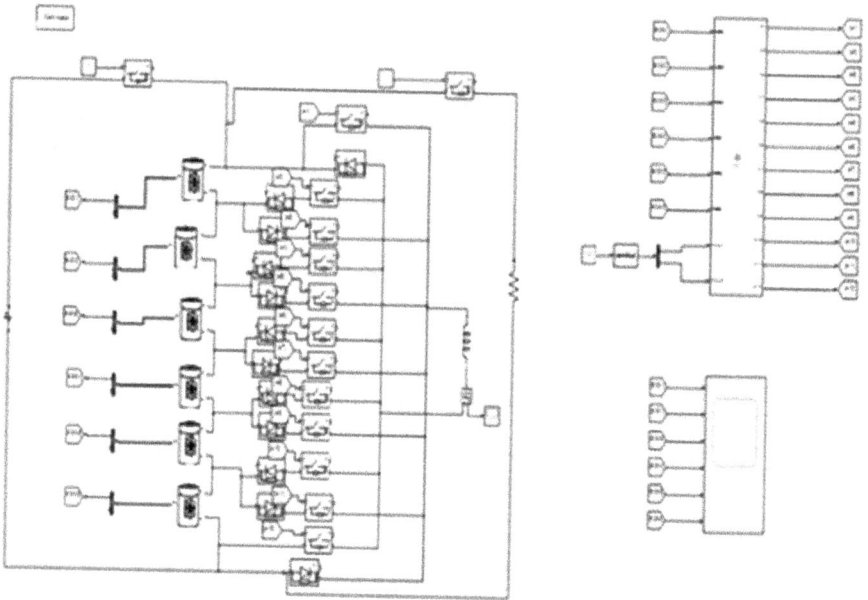

FIGURE 7.5
MATLAB circuit.

TABLE 7.1

State of Charge of All Cells

Cell Number	State of Charge
Cell 1	41
Cell 2	40
Cell 3	44
Cell 4	43
Cell 5	47
Cell 6	46

7.7 Result and Observations

The following various values, as indicated in Table 7.1, are entered into the MATLAB circuit. The results for the different cases—Static, Charging, and Discharging—are as follows.

7.7.1 Static Mode

Figure 7.6 shows that balancing is completed in approximately 77 seconds, which is a good speed because it shows that the SOC of cells 6 and 5 balanced faster than those of cells 1 and 2, despite the fact that they both had the same difference. This is because cell 6 starts to lose energy faster when it discharges through the inductor, which speeds up the process and causes cells 6 and 5 to balance quickly, whereas cell 1 requires an inductor, which takes more time, which is why it took longer.

7.7.2 Charging

MATLAB®/Simulink® results in charging mode are displayed in Figure 7.7. It takes nearly seven seconds to balance every cell. Subsequently, every cell begins to charge uniformly. Additionally, it prevents the batteries from over-charging. Because less-charged cells are being charged alongside other cells by an external source, balancing occurs faster during charging mode than during static mode.

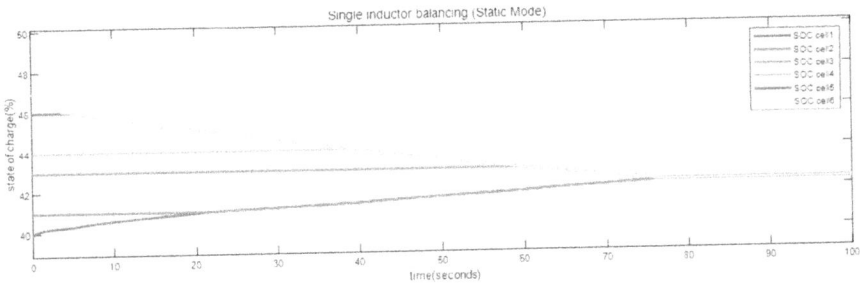

FIGURE 7.6
Simulink results (static).

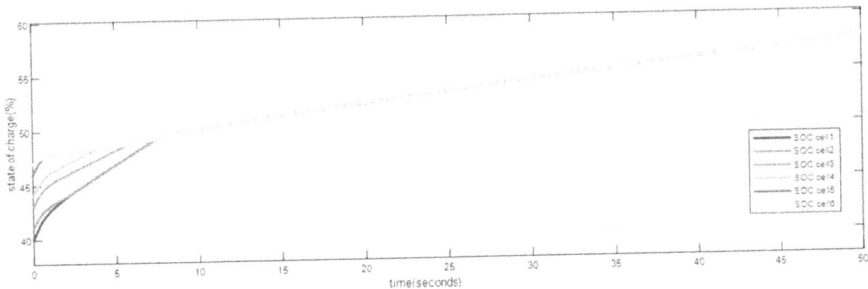

FIGURE 7.7
Simulink results (charging).

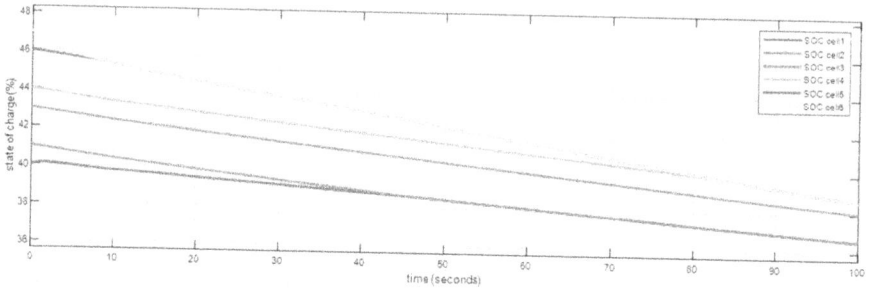

FIGURE 7.8
Simulink results (discharging).

7.7.3 Discharging

MATLAB findings for active cell balancing in discharging mode are displayed in Figure 7.8. Here, the battery pack is linked to a 5-ohm resistive load, and it took about 165 seconds for all the cells to balance. When compared to the other two modes, the results obtained in this method are the slowest.

7.8 Conclusions

The single-inductor active cell-balancing circuit with an automated function controller is the focus of this work. The circuit in question is capable of operating in all three modes—static, charging, and discharging. The circuit operates most quickly when it is charging, taking only 7 seconds to balance the entire pack of six batteries. When it is discharging, the circuit takes 165 seconds to balance. MATLAB Simulink is utilized to obtain all of the results.

7.9 Future Aspects

An important future direction for active cell balancing methods is the integration of artificial intelligence (AI) and machine learning. These technologies can significantly improve performance by forecasting cell behavior, optimizing the order and timing of balancing, and adjusting control strategies based on real-time data such as temperature, voltage variations, and cell aging. AI-enabled battery management systems (BMS) can efficiently oversee balancing processes to enhance both reliability and energy efficiency, while also reducing energy loss and cell degradation. Furthermore, progress in wireless communication and modular BMS design is expected to support more adaptable and intelligent balancing solutions. In summary, although passive balancing will continue to be used in simpler and cost-sensitive applications, active balancing—augmented by AI and advanced control techniques—is set to become a key component in the future of large-scale, high-performance energy storage systems.

References

Daowd, M., Omar, N., Van Den Bossche, P., & Van Mierlo, J. (2011). Passive and active battery balancing comparison based on MATLAB simulation. In *7th IEEE Vehicle Power and Propulsion Conference, VPPC'11* (pp. 1–7). IEEE. https://doi.org/10.1109/VPPC.2011.6043010

Duraisamy, T., & Kaliyaperumal, D. (2020). Active cell balancing for electric vehicle battery management system. *International Journal of Power Electronics and Drive Systems, 11*(2), 571.

Duraisamy, T., & Kaliyaperumal, D. (2021). Machine learning-based optimal cell balancing mechanism for electric vehicle battery management system. *IEEE Access, 9*, 132846–132861. https://doi.org/10.1109/ACCESS.2021.3115255

Marcin, D., Lacko, M., Bodnár, D., & Pancurák, L. (2024). Capacitor-based active cell balancing for electric vehicle battery systems: Insights from simulations. *Power Electronics and Drives, 9*(1), 317–330. https://doi.org/10.2478/pead-2024-0020

Mohamed Dawod, M., Omar, N., Van Den Bossche, P., & Van Mierlo, J. (2011). Passive and active battery balancing comparison based on MATLAB simulation. In *7th IEEE Vehicle Power and Propulsion Conference, VPPC'11* (pp. 1–7). IEEE, Chicago, Illinois, USA.

Safitri, W. Y., Suryoatmojo, H., & Lystianingrum, V. (2022). Balancing control strategy of lithium-ion using proportional derivative-fuzzy logic controller. In *2022 5th International Seminar on Research of Information Technology and Intelligent Systems (ISRITI)* (pp. 520–524). IEEE. https://doi.org/10.1109/ISRITI56927.2022.10052829

Shah, S., Murali, M., & Gandhi, P. (2018). A practical approach of active cell balancing in a battery management system. In *2018 IEEE International Conference on Power Electronics, Drives and Energy Systems (PEDES)* (pp. 1–6). IEEE, Madras, India.

Vardhan, R. K., Selvathai, T., Reginald, R., Sivakumar, P., & Sundaresh, S. (2017). Modeling of single inductor based battery balancing circuit for hybrid electric vehicles. In *IECON 2017–43rd Annual Conference of the IEEE Industrial Electronics Society* (pp. 2293–2298). IEEE, Beijing, China.

8

Integrated Simulation and Hardware-in-the-Loop Analysis of ISO15118-Based Smart EV Charging Communication Systems

Bhavesh Karnik, Arigela Satya Veerendra, and Kumaran Kadirgama

8.1 Introduction

Today, the need for electric vehicles makes India one of the top ten car markets in the world, with a rapidly growing middle class, purchasing power, and stable economic growth (global-ev-outlook-2021). However, in the previous 2 years, gas prices have increased by more than 50% in 13 separate cases. India may experience demand for alternative vehicle technologies, including EVs, in the future. Although the initial investment is about 1.5 times that of a conventional IC engine, environmental costs have become a bigger issue than car costs. The national government focuses on research and development, while municipalities support the implementation of local infrastructure through public-private partnerships and other programs. Charging infrastructure will play a critical role in EV deployment, and without a proactive strategy and timeline, it is a major barrier to widespread market adoption. Charging infrastructure includes all the technology and software needed to deliver electricity from the grid to cars. It can be classified by location, power level, and charging time method (Advances in Electronic and Electric Engineering (ripublication.com)). With the rapid development of the EV industry, EV chargers have become an important infrastructure to promote the development of EVs. The rise of electric car charging stations provides convenience and flexibility to electric vehicle owners, while also benefiting the environment, energy prices, and sustainable development. This article analyzes the benefits of EV chargers and the advantages of smart EV chargers. Not only is it a simple charging alternative for car users, but it also helps reduce environmental pollution and energy costs. Electric car charging stations play an important role in the development of sustainable transportation.

DOI: 10.1201/9781003506478-8

Advantages of EV chargers. Car owners can avoid constantly searching for gas stations by using EV chargers that provide convenient charging alternatives. In addition, a larger network of charging stations expands the range of charging options. Conventional gasoline cars are being replaced by EVs that reduce emissions. The carbon footprint of EV chargers is reduced due to the use of renewable energy (www.vector.com/int/en/know-how/smart-charging/#c237833).

Smart charging refers to the charging process of electric or hybrid cars based on ISO 15118, DIN SPEC 70121, and SAE J2847/2 standards. There are two ways for vehicles and charging stations to communicate:

- Powerline communication, achieved through control test points using Pulse Width Modulation (PWM) signals and digital signals as specified in the HomePlug-GreenPhy standard.
- Wireless communication, when using inductive charging.

The main goal is to reduce emissions, and the compatibility of all vehicles with charging infrastructure from different providers is a key element in ensuring the rapid success of electromobility. In addition, this article seeks to improve the sustainability of transportation by encouraging the adoption of electric vehicles and improving EV charging infrastructure. Communication protocols between EVs and EVSEs are essential for safe, efficient, and cost-effective charging.

The charging process is more complicated than it first appears, especially when considering the transition to alternative energy sources and the upcoming smart grid. Not only basic details such as available power at the charging station, battery health, and charging demand must be considered. Therefore, smart charging according to ISO 15118 will prove itself in the future. Communication between vehicles and charging stations, as well as the distribution and supply of electrical energy, is one of the main challenges. Thus, in this paper, we present the communication and analysis of EVSE and EV.

8.2 Recent Studies

The following literature review gives detailed information about previous work in the field of smart charging. The EV-EVSE communication is implemented with the Open Charge Point Protocol as well as Ethernet instead of PLC using MATLAB and SIMULINK. The simulator further enhances the charging protocol (Madhusudhanan & Sivraj, 2022). Voltage violations can result from the uncontrolled charging of an EV. Consequently, communication and information technology emerge as highly viable alternatives for the integration of EVs into the electrical grid, thereby converting their adverse effects into a managerial opportunity for the smart grid. ISO/IEC 15118, a worldwide standard for communication between the EV and the charging station, is critical in this regard. This allows for the exchange of energy-relevant information and, as a result, a more flexible charging and regenerative process (Berrada et al., 2021). A conventional technique to mitigate this risk is to organize tests between all implementations to explicitly test compatibility. This strategy, however, is hard and costly since it necessitates substantial testing and coordination among all implementers. It specifies compliance tests that can be completely automated. This is a novel approach to machine-to-machine interface requirements that includes not just communication but also power interfaces (Hänsch et al., 2014). It looks at communication methods and refers to suitable electric vehicle standards. Different charging modes enable the charging of an electric car, when a succession of complicated circumstances is desirable, as well as AC and DC charging with high power levels. The direct current communication technique is explained and analyzed in conjunction with the ISO 15118 standard, along with its structure. It also explains the OCCP protocol, which provides communication between an operator and a charger (Parchomiuk et al., 2019). This study considers that noise coupling in the cable bundle is the primary cause of communication performance loss. The noise originates from the EVSE power conversion equipment. To examine the association between produced noise and coupling noise in communication systems, it's important to consider the obvious service failure and the presence of coupled noise in the prescribed frequency range (Park et al., 2023). This study suggests monitoring ISO 15118 communication between electric cars using the V2G (Vehicle-to-Grid) sniffer system and Wireshark. It aims to understand the communication between SECCs and EVCCs using the ISO 15118 protocol. The V2G sniffer system consists of a Nucleo sniffer board that intercepts messages from the EV or supply equipment utilizing UART. These messages are then forwarded to the PC via Ethernet and analyzed with Wireshark, allowing for real-time communication data capture. This research analyzes ISO 15118 communication patterns, message exchanges, and improvement options (Mültin, 2018). A smart charging approach for a PEV network provides several charging alternatives, such as AC level 2 charging, DC rapid charging, as well as battery swapping at charging stations. With the aim of reducing charging time, journey time, and cost, it characterizes the ideal PEV charging station as a multi-objective optimization problem. It proposes a metaheuristic approach based on ant colony optimization. Simulation findings indicate that the suggested technique considerably lowers waiting time and charge costs (Moghaddam et al., 2017). According to the literature review, the implementation and operation of charging infrastructure require EV-EVSE communication. There has been little research on ISO 15118 in simulation for communication standards. Most of the simulations available are based on DIN/IEC standards or the Open Charge Point Protocol, and their work on EV-EVSE connection is therefore quite restricted. However, combining hardware and software simulators enables greater flexibility in the inclusion of various functions and allows for the easy integration of different technologies. A combination-based simulation with the features required to execute both EV-EVSE simulation and its security aspects would be more useful in comprehending these communications.

8.3 Methodology

8.3.1 Methodology Adopted

The following topics, EV-EVSE with the VT system, will be simulated in real time by a combination of hardware and software. However, CANoe guidelines for simulation must be followed. CAPL programming is essential for writing test cases to be implemented in vTESTstudio. Based on the basics and considering the project requirements, CAN, Ethernet, and smart charging protocols should be analyzed and understood. Devices such as VT systems and smart charge cards must be in the loop with Vector devices such as the VN5620, which must be modified to access the main thread. After the inter-system connection, the ISO 15118 protocol will be used in CANoe version 17 SP3. After the process, EV and EVSE test packages are offered. With all these Vector tools, EV-EVSE simulation with the VT system is simulated.

Figure 8.1 gives brief information about the block diagram of the complete system. The system is connected to the VN5620 hardware via USB. The VT

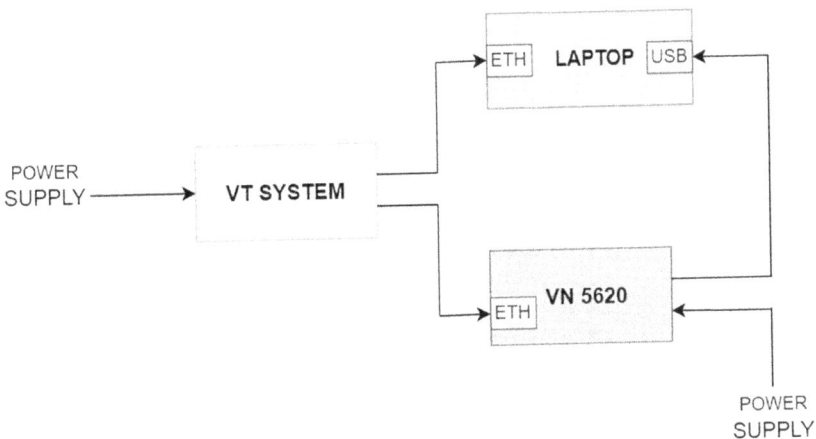

FIGURE 8.1
Simulation of block diagram of EV_EVSE system.

system is attached via an Ethernet cable to the VN hardware, and the connection on the other side of the VT is made with Ethernet to the extended screen of the system.

Figures 8.2 and 8.3 explains the smart charging communication between an EV and EVSE in the CANoe simulation in the "real mode."

Figure 8.4 depicts the EV-EVSE simulation, in which all panels are built in CANoe and the EV and EVSE are linked to their respective VT cards. The protocol panel performs the SLAC procedure, resulting in a complete simulation.

Figure 8.5 explains that the CANoe Test Package EV comprises a collection of robust test cases. This is utilized to test electric vehicle (EV) interoperability as well as conformance with the international standards CHAdeMO, NACS (ISO 15118 and DIN 70121), CCS (ISO 15118 and DIN 70121), and GB/T. The relevant test cases are covered by the CANoe Test Package EV, which is properly organized and updated regularly (www.vector.com/int/en/search/search-overview/#q=Test%20package).

Figure 8.6 states that the CANoe Test Package EVSE comprises a collection of robust test cases. This is utilized to test EVSE for interoperability as well as compliance with the international standards GB/T and CCS (ISO 15118).

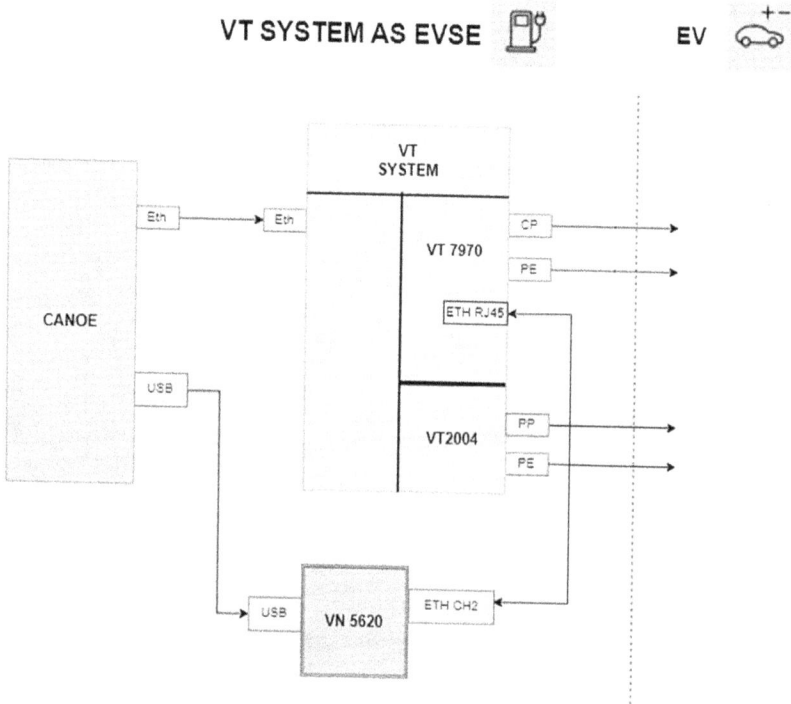

FIGURE 8.2
Circuit diagram of VT_EVSE_7970 ("VT System for Smart Charge Communication," Vector).

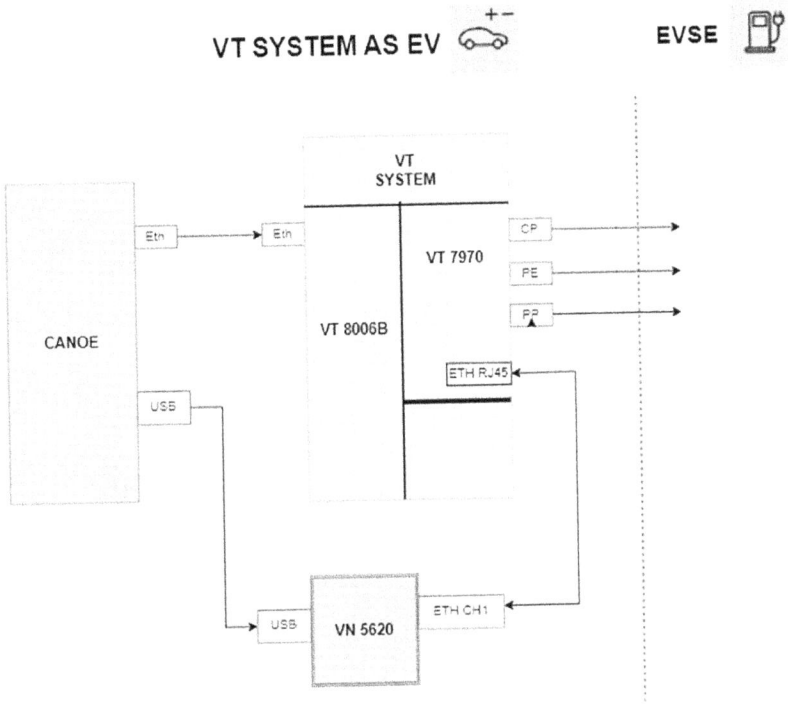

FIGURE 8.3
Circuit diagram of VT_EV_7970 ("VT System for Smart Charge Communication," Vector).

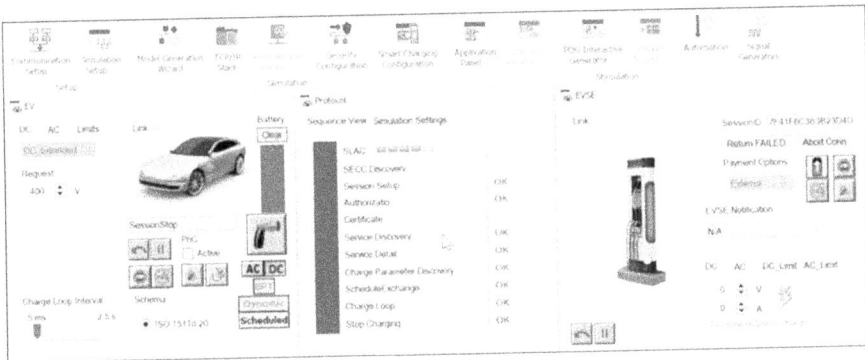

FIGURE 8.4
EV-EVSE configuration in CANoe17 SP3.

The CANoe Test Package EVSE is neatly organized, covers pertinent test situations, and is updated often (www.vector.com/int/en/search/search-overview/#q=Test%20package).

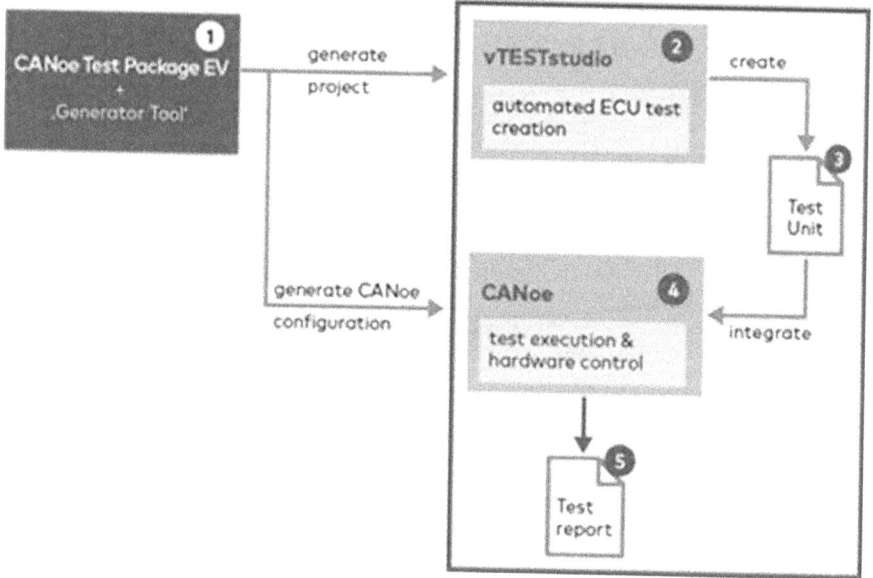

FIGURE 8.5
EV test package (www.vector.com/int/en/search/search-overview/#q=Test%20package).

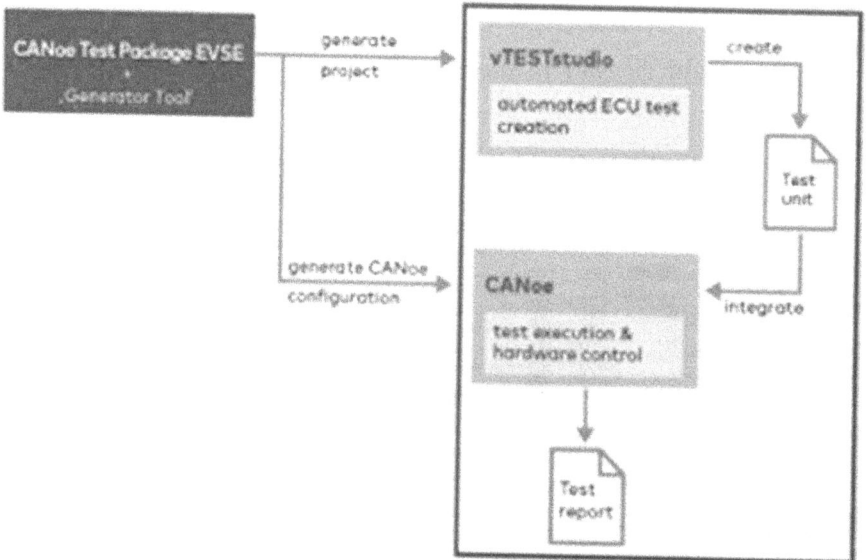

FIGURE 8.6
EVSE test package (www.vector.com/int/en/search/search-overview/#q=Test%20package).

8.3.2 Tools Required

- **CANoe**: For the development, analysis, and testing of individual ECUs or complete networks of ECUs, CANoe is considered a versatile tool. Throughout the whole development process, from planning to the start of distributed systems or individual ECUs, it assists network designers, as well as development, and test engineers at OEMs and suppliers. During the first stages of the development process, CANoe is utilized to generate simulation models that accurately replicate the actions of the Electronic Control Units (ECUs) (Bozdal et al., 2018). Throughout the ongoing ECU development process, these models are used as the foundation for the analysis, testing, as well as integration of bus systems and ECUs. This enables the early detection and subsequent correction of issues. The software offers graphical as well as text-based analysis windows for assessing the outcomes. CANoe includes the Test Feature Set, which facilitates and automates the execution of tests. This tool is used for modeling and executing sequential test sequences, as well as automatically generating a comprehensive test report. The Diagnostic Feature Set is accessible in CANoe for establishing diagnostic connections with the Electronic Control Unit (ECU) (www.vector.com/int/en/download/fact-sheet-canoe/).

- **vTESTstudio**: For developing automated ECU testing, vTESTstudio offers a potent development environment. It offers programming-based, table-based, and graphical test notations and test development approaches to improve test design efficiency and streamline reusability. All stages of the product development process, from model testing to system validation, may make use of vTESTstudio. Because it uses an open interface, vTESTstudio is easily integrated into tool landscapes that already exist. The ability to separate test logic, implementation, parameter values, and test vectors makes it simple to reuse. Event preprocessing in C# and CAPL may be used for test sequence access and assessment. A vTESTstudio project comprises an arbitrary number of test units. In CANoe, a test unit refers to the unit that may be configured and executed. The test unit comprises a collection of files, including CAPL files, C# files, test tables, test diagrams, and parameter files. A vTESTstudio project encompasses the testing of an Electronic Control Unit (ECU), whereas a test unit is responsible for implementing tests for a specific feature of the system being tested. vTESTstudio offers many components that contribute to the creation of a productive testing environment, including the consolidation of graphical test design editors, programming editors, parameter editors, and curve editors into a single tool. The Test Diagram Editor allows for explicit modeling of the test logic, while the State Diagram Editor enables representation of the anticipated behavior of the SUT. This provides convenient capabilities for conducting reviews. Reusability is facilitated by segregating test logic, implementation, parameter values, and test vectors. Events can be processed in CAPL and C# to facilitate access and assessment in test sequences. The tool also offers enhanced support for incorporating different versions of test structures, implementation, and parameters into a unified system. Strategies are available for achieving comprehensive test coverage and ensuring the ability to track and trace externally specified requirements as well as test descriptions throughout the test implementation and in the test report. This automatically created documentation serves the purpose of test design and is intended for both internal and external reviews (www.vector. com/int/en/products/products-a-z/software/vteststudio/).

VT System: A hardware device that may be added or removed to regulate ECU I/O connections for testing is called the VT device. The tests are coded in Vector CANoe, and the connections are managed using CANoe. The VT System modules are directly connected to the actuator and sensor connections of the ECU that must be tested. The VT modules may also be linked to the original actuators and sensors. On the other hand, the VT modules may be used to imitate them. The measured and preprocessed output signals from the ECU are sent in processed form to the CANoe test programs. The CANoe test program may specify the stimulation signals that are generated on the VT module for the inputs of the ECU. Several VT System modules feature a user-programmable FPGA, which allows for the creation of specific stimulation signals and personalized processing of the measured signals. A variety of electrical problems, such as short circuits between ECU lines, line breakage, or short circuits to ground or Vbatt, may also be produced by the VT modules. The VT modules are installed into one or more 19 racks that include a backplane as part of the VT System. The lowest part of the back is occupied by the backplane, while the top section provides direct access to the module connections. These connections are directly inserted into the original loads and the ECU lines. CANoe is connected via an Ethernet cable using a special, real-time-capable industrial Ethernet protocol. The backplane links the EtherCAT bus and the power supply to the inserted VT modules. No specific PC hardware is required for the PC running CANoe; all it needs is an Ethernet connection ("VT System for Smart Charge Communication," Vector).

- **Smart Charging Module VT 7970**: A specialized module called the VT7970 is used to evaluate electric vehicle smart charge communication. It is the combination of a VT7900A and an application board that is mounted on the VT7900A. The module can simulate both communication partners, the EVSE and the EV itself, and offers the following features: Control pilot (CP) circuit for PWM communication according to IEC 61851-1 Annex A, electrically isolated from the remaining VT System Power line communication (PLC) with dLAN, GreenPHY module, which is integrated on the VT7870 and communicates with CANoe via an RJ45 connector. Voltage measurement of proximity contact. Several possibilities exist to simulate errors and varying component values. PWM signal will be generated and measured on the application board, but external measurement and stimulation are also possible Parameters of the PWM signal and the relays to switch the signal paths can be controlled in CANoe via system variables (VT7970 – Smart Charging Communication Module for the VT System; "Technical papers on the development of Embedded Electronics," Vector- Automotive).

- **VN 5620**: The VN5620 is a powerful and compact interface specially designed for examination, testing, simulation, and analysis of Ethernet networks. The VN5620 interface can accommodate a variety of applications. Network participation is suitable for synchronous control of Ethernet and other bus systems. The user interface to the computer can be USB 3.0 or Ethernet (1000BASE-T) (www.vector. com/int/en/products/products-a-z/hardware/network-interfaces/vn5620-vn5430/).

- **CANoe EV Test Package**: The CANoe EV Test Package contains a powerful collection of test cases. It is used to determine whether electric vehicles meet or comply with the requirements of international standards GB/T, CHAdeMO, NACS, and CCS (ISO 15118 and DIN 70121). The CANoe Test Package is well organized, includes relevant test cases, and is updated regularly. CANoe and vTESTstudio are part of the toolkit, which includes the execution and modification of test cases. Thanks to the simple operation of the device and the seamless tool chain, reliable test results can be achieved very efficiently. Test case execution can be performed at different development stages depending on the hardware used, which can also be provided by third-party providers.

    Tool Functionality: The vTESTstudio project is used to implement the test cases generated by the Canoe Test Package EV. The actual test execution takes place in CANoe. When a vTESTstudio project is created, the generator tool is used to generate a new CANoe configuration with ready-to-run test cases. The test unit is generated after modifications are made to the vTESTstudio project and then added to CANoe. In CANoe, the required test cases can be selected one by one or in groups and run immediately. The test report contains the test results (www.vector.com/int/en/search/search-overview/#q=Test%20package).

- **CANoe EVSE Test Package**: CANoe EVSE Test Package is a collection of reliable test cases. It is used to assess the conformity of electric vehicle supply equipment (EVSE) to international standards CCS and GB/T. The CANoe Test Package EVSE includes relevant test cases, is regularly updated, and regularly built. Efficiency and compatibility tests are carried out automatically according to GB/T, CCS, and CHAdeMO charging standards.

How the Tool Works:

Test cases generated by the CANoe Test Package EVSE are executed as projects in vTESTstudio. Test execution is performed in CANoe. The "Generator Tool" is used to generate a new CANoe configuration containing test cases prepared for execution at the same time as the vTESTstudio project is created. After modifying the WTESTstudio project, a unit test was created and then added to CANoe. Test cases can be selected individually or as a CANoe group and started immediately. The test results are recorded in the test report (www.vector.com/int/en/search/search-overview/#q=Test%20package).

8.4 Results and Analysis

The above two EV-EVSE simulation results are created graphically and are synced with the messages in the trace window, as shown in Figure 8.7, and the data are represented in Table 8.1, respectively. The graph depicts a

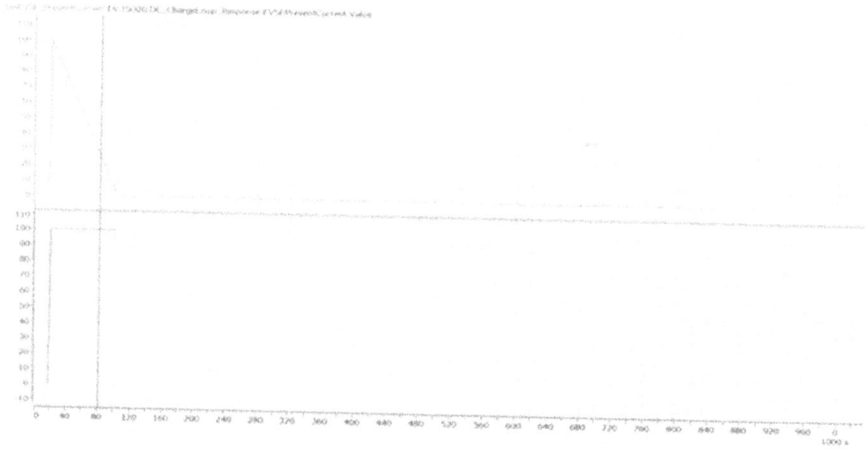

FIGURE 8.7

Graphical analysis of EV-EVSE in CANoe 17 SP3.

TABLE 8.1

Trace Window of EV-EVSE

Time	Chn	Port(s)	VLAN	Sim	Dir	Protocol	Sender	Node	Receive
42.413394	Eth	1	EV_Node	s	Rx	tls	Application	Data	122
42.413396	Eth	1	EVSE_Node		Tx	tls	Application	Data	122
42.463398	Eth	1	EVSE_Node	s	Rx	tls	Application	Data	137
42.4634	Eth	1	EV_Node		Tx	tls	Application	Data	137
42.513402	Eth	1	EV_Node	s	Rx	tls	Application	Data	122
42.513403	Eth	1	EVSE_Node		Tx	tls	Application	Data	122
42.563405	Eth	1	EVSE_Node	s	Rx	tls	Application	Data	137
42.563407	Eth	1	EV_Node		Tx	tls	Application	Data	137
42.613409	Eth	1	EV_Node	s	Rx	tls	Application	Data	122
42.613411	Eth	1	EVSE_Node		Tx	tls	Application	Data	122
42.663413	Eth	1	EVSE_Node	s	Rx	tls	Application	Data	137
42.663415	Eth	1	EV_Node		Tx	tls	Application	Data	137
42.713417	Eth	1	EV_Node	s	Rx	tls	Application	Data	122
42.713419	Eth	1	EVSE_Node		Tx	tls	Application	Data	122
42.763421	Eth	1	EVSE_Node	s	Rx	tls	Application	Data	137
42.763423	Eth	1	EV_Node		Tx	tls	Application	Data	137
42.813425	Eth	1	EV_Node	s	Rx	tls	Application	Data	122
42.813426	Eth	1	EVSE_Node		Tx	tls	Application	Data	122
42.863428	Eth	1	EVSE_Node	s	Rx	tls	Application	Data	137
42.86343	Eth	1	EV_Node		Tx	tls	Application	Data	137

study of the current of an EV and an EVSE, as illustrated in Figure 8.8. The EVSE is fully charged and then returns to its previous position, whereas the EV displays an increase in charging until it reaches 100% and then stabilizes.

8.5 Conclusion and Future Scope

The communication between the EV and EVSE is established, and the charging gun is attached to the EV. After stable communication, the SLAC process is initiated, and a specific current of 30 amps is set. Once the voltage matches, the current is set to a certain ampere, and charging takes place. When the EV is completely charged, the process between the EV and EVSE is completed. The configuration of the EV-EVSE is simulated and analyzed according to specific charging standards, and the above results are obtained after the combination of hardware and software is carried out in a proper loop.

This simulation of EV-EVSE charging in ISO 15118 with Ethernet over OCPP (Madhusudhanan & Sivraj, 2022) is a major factor that brought a new idea of

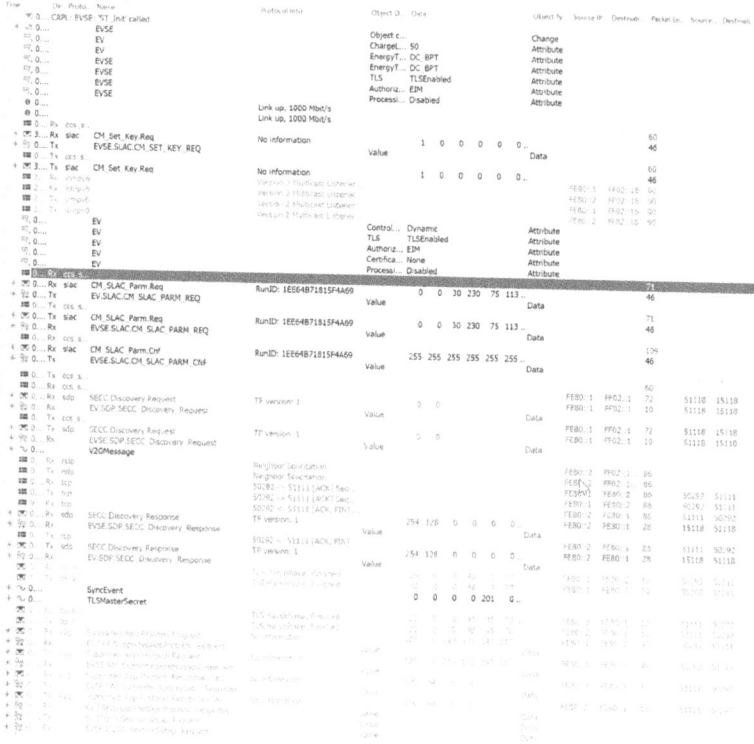

FIGURE 8.8
Trace window of EV-EVSE in CANoe 17 SP3.

PLC communication with the help of Vector devices. Even with the same simulation, adding the concept of security to the configuration will help EV and EVSE manufacturers. If there is no security addition, the exchange of data between the EV and EVSE can be altered, resulting in data loss. So, the Transport Layer Security plays an important role in securing the data, which is an upcoming scope in the field of smart charging.

References

AEEE, Advances in Electronic and Electric Engineering, Journals Publishers, Science Journal Publisher in India, Indian Subscription Agency, Indian Books Distributors (ripublication.com) [Accessed on 10th May 2024].

Berrada, A., Annen, F., Gurcke, M., & Haubrock, J. (2021, June). Integrating electric vehicle communication in smart grids. In *2021 IEEE Madrid Power Tech* (pp. 1–5). IEEE, Madrid, Spain.

Bozdal, M., Samie, M., & Jennions, I. (2018, August). A survey on can bus protocol: Attacks, challenges, and potential solutions. In *2018 International Conference on Computing, Electronics & Communications Engineering (iCCECE)* (pp. 201–205). IEEE, University of Essex, Southend, UK.

Hänsch, K., Pelzer, A., Komarnicki, P., Gröning, S., Schmutzler, J., Wietfeld, C., ... & Müller, R. (2014, July). An ISO/IEC 15118 conformance testing system architecture. In *2014 IEEE PES General Meeting | Conference & Exposition* (pp. 1–5). IEEE, Maryland, USA.

https://www.iea.org/reports/global-ev-outlook-2021/policies-to-promote-electric-vehicle-deployment [Accessed on 10th May 2024].

https://www.vector.com/int/en/download/fact-sheet-canoe/ [Accessed on 4th May 2024].

https://www.vector.com/int/en/know-how/smart-charging/#c237833 [Accessed on 2nd May 2024].

https://www.vector.com/int/en/products/products-a-z/hardware/network-interfaces/vn5620-vn5430/ [Accessed on 13th May 2024].

https://www.vector.com/int/en/products/products-a-z/software/vteststudio/ [Accessed on 10th May 2024].

https://www.vector.com/int/en/search/search-overview/#q=Test%20package [Accessed on 4th May 2024].

Madhusudhanan, S., & Sivraj, P. (2022, November). Development of communication simulator for electric vehicle charging following ISO 15118. In *2022 IEEE North Karnataka Subsection Flagship International Conference (NKCon)* (pp. 1–6). IEEE, Vijayapura, India.

Moghaddam, Z., Ahmad, I., Habibi, D., & Phung, Q. V. (2017). Smart charging strategy for electric vehicle charging stations. *IEEE Transactions on Transportation Electrification*, 4(1), 76–88.

Mültin, M. (2018, July). ISO 15118 as the enabler of vehicle-to-grid applications. In *2018 International Conference of Electrical and Electronic Technologies for Automotive* (pp. 1–6). IEEE, Milan, Italy.

Parchomiuk, M., Moradewicz, A., & Gawiński, H. (2019). An overview of electric vehicles fast charging infrastructure. *2019 Progress in Applied Electrical Engineering (PAEE)*, pp. 1–5.

Park, S., Lee, E., Noh, Y. H., Choi, D. H., & Yook, J. G. (2023). Accurate modeling of CCS combo type 1 cable and its communication performance analysis for high-speed EV-EVSE charging system. *Energies*, 16(16), 5947.

"Technical Papers on the Development of Embedded Electronics", Vector- Automotive, 7th Edition, [Accessed on 13th May 2024].

"VT System for Smart Charge Communication", Vector. [Handbook accessed on 30th April 2024].

VT7970 – Smart Charging Communication Module for the VT System | Vector [Accessed on 11th May 2024].

9

AI-Driven Analysis of Air Gap Eccentricity Effects in PMSM for Condition Monitoring in Smart Industrial Systems

Yashoda Asangihal, B. L. Rajalakshmi Samaga, and Babul Salam Ksm Kader Ibrahim

9.1 Introduction

Due to the clear advantages of having a simple and compact structure, being easy to manufacture, having high power density, and being highly efficient, permanent magnet machines have found extensive use in a wide range of industries, including military, wind turbines, electric vehicles (EV), railways, and military aircraft. As a result, they have drawn more attention from the academic and business communities. Various PM machine topologies include stator-PM, axial-field, double-stator, flux memory, and transverse flux machines. Multiport, fault-tolerant, and multiphase PM machines are examples of PM machines. In certain applications, such as electric vehicles and aircraft, it is essential that the PM machine be dependable and secure.

The PM machine is prone to malfunctions because of its high-power density, complicated working circumstances, compact installation area, and unfavorable temperature environment. In addition to affecting the PM machine's performance, the flaws may cause harm or even catastrophic system breakdowns. In order to ensure fault-tolerant operation, a variety of control techniques for PM machines under fault situations have been devised. It is obvious that PMSM fault diagnosis is essential for both scheduling maintenance to increase the overall system's dependability and security as well as for ensuring fault-tolerant operation. Air pollution has become a global issue in addition to economic ones; thus, governments and academic institutions have been generating. It has been acknowledged that EVs will be the driver of the automotive industry's growth and change in the twenty-first century. EV production relies on several important technologies, including the motor drive system, body design, energy system, and management and system-level optimization. In order to increase a vehicle's performance on

DOI: 10.1201/9781003506478-9

the road, aerodynamic drag will be considered when designing the body. The motor drive system, comprising three primary parts (motor, power converter, and control), is a crucial component, as it is a power conversion unit that greatly influences electric vehicle performance. Although the DC motor drive was first designed for electric vehicles, it needs routine maintenance for its brushes and commutator. Nevertheless, as permanent magnet technology, computer technologies, and motor drive control engineering have advanced, PMSM AC synchronous motors have very high efficiency and reliability, and are brushless. Permanent magnets in the rotor portion and three-phase outputs from the inverter supply the stator part of the PMSM with field excitation. Permanent magnets are the primary cause of the rotor part difference between PMSM and induction motors. A hybrid of an induction motor and a BLDC motor is the PMSM. Furthermore, similar to an induction motor, the machine has an air gap and a stator construction with windings producing flux density sinusoidal in nature. The rotor of a PMSM motor contains permanent magnets, and the motor is structurally synchronous. Permanent magnets are positioned.

The outer rotor SPMSM can be modeled using any one of three methods:

- dq modeling
- Multiple coupled circuit approach
- Finite Element Analysis (FEA)

The dq model assumes that the air gap flux remains the same over the circumferential length in the air gap. Hence, it is not advisable to use the dq modeling method to study the performance of a motor under faulty conditions [1]. The multiple coupled circuit approach [2,3] can incorporate the variation in the flux due to a mechanical fault, but it demands extensive knowledge of the effect of the fault on flux. In FEA physical model of the machine can be developed, and any change in the physical parameters can be incorporated very easily. Hence, FEA is adopted to develop the model under the air gap eccentricity condition.

9.2 Literature Review

Florina-Ambrozia's paper focuses on simulating a PMSM, which is meant to supply electricity to a scroll compressor of an air conditioning unit in automobiles. The simulations were necessary for both the PMSM's electromagnetic behavior study and the pre-sizing verification of the PMSM, which is done with the aid of simulation. The ANSYS Motor-CAD program is utilized to complete the model. All of the data showed that the PMSM pre-sizing was done appropriately and that the electrical machine that was designed could meet all of the requirements [4]. Bdewi's paper looks into a potential way to increase the PMSM utilized in EVs' torque density. Other machine specifications were also considered and maintained within a reasonable range at the same time. This was accomplished by implementing performance-enhancing techniques, such as adjusting the machine's size for best performance and researching high-efficiency winding topologies. The suggested design was implemented using the Magnet 7.4.1 software package, including static and transient finite element technique solvers [5]. In Ayhan's paper, Doğukan presents an inner rotor PMSM architecture with radial flux for use in electric car applications. The ANSYS Motor-CAD software will be utilized to analyze the Interior PMSM of the Nissan Leaf EV model, manufactured in early 2012. The results of the analysis will be cross-checked with theoretical findings before the motor is optimized for the project. This research aims to shed light on different parameter changes and how they affect other parameters [6]. In Viswanath's work, the dynamic model of an induction motor for static and transient 2D analysis is created using the CAD program Magnet. The numerous parameters of the machine, including stator current, magnetic torque, and magnetic flux density, are calculated and compared under both favourable and unfavourable conditions [7].

The modeling and diagnostics of PMSM with respect to all spatial harmonics is the focus of Gherabi's work. To do this, a winding function-based mathematical model was developed based on two scenarios: one in which the machine may have an eccentricity fault, and the other in which the air gap is constant. [8]. Manel Krichen's research paper presents the impact of static eccentricity defects in PMSMs on flux linkage, air gap flux density, electromagnetic torque (EM torque), and electromotive forces (EMFs). When modeling PMSM, a 2D finite element method (FEM) must be utilized in both ideal and imperfect circumstances [9]. The research on PMSM operating with eccentricity and bearing damage is presented in Rosero's paper. The goal is to use the existing signature analysis to find and identify the defect [10]. In Wang, Lingling's study, the torque of an electric vehicle's PMSM is analyzed using finite element analysis. First, the PMSM is modeled in two dimensions. The model's accuracy can be confirmed by looking at the flux density distribution and magnetic field lines of the PMSM under no load. Second, an analysis is done on the PMSM's torque variation curve at rated load. It demonstrates that the motor's torque satisfies the stability requirements. In the end, one can ascertain that motor torque and current amplitude can be related non-linearly by varying the current stimulation and analyzing the torque shift of the motor. This provides a basis for enhancing the performance of the PMSM found in EVs. [11]. According to Tikadar's research, a co-simulation method utilising MATLAB/Simulink and Ansys software (Maxwell and Twin Builder) is demonstrated for the PMSM model. This technique can enhance the PMSM's design and assess its performance using Rotating Machine Expert (Rampart) in the event that any slight changes to the parameters result in output inaccuracy [12]. The effects of speed, forced convection heat transfer coefficient, and current density on the PMSM's maximum temperature, overall efficiency, and heat losses have all been thoroughly examined. Ultimately, an efficiency map derived from the linked electromagnetic simulation has been deciphered [13].

According to AKYÜN, Yakup research, using Ansys Maxwell 2D software, an IPMSM is analytically developed, analyzed, and built at Bader Motor Technology Company. According to the specifications provided by the customer, the motor has 0.5 Arms, 0.2 Nm, and 3,400 rpm in its design [14].

9.3 Methodology

9.3.1 Development of Healthy PMSM Model

ANSYS Motor-CAD is used in the development of the PMSM model. The software's electromagnetic module, dubbed E-Magnetic, is used to examine the electromagnetic behavior of the created PMSM model. This makes it possible to calculate the electromagnetic performance using FEA techniques in addition to accurate analytical techniques. Eight preconfigured electromagnetic templates are provided by the software to the designers for use in the analysis of the BPM category. From the motor type menu, we can choose the required motor for the simulation. The model is developed using the steps listed below.

9.3.1.1 Step 1: Implementation

The primary dimensions of the iron cores (air gap length, shaft diameter, exterior and interior stator diameters, length, number of stator teeth, and slots) need to be determined in the boxes of the two panels. The software's Geometry menu and the created stator and rotor iron cores are used to set the diameters of the PMSM's axial and transversal cross sections. Parameters such as stator parameters, axial parameters, pole parameters, and radial parameters of the PMSM motor are listed in Tables 9.1–9.4, respectively. The outer rotor Surface Permanent Magnet (SPM) synchronous motor structure and the inner rotor Surface Permanent Magnet (SPM) synchronous motor structure are shown in Figures 9.1 and 9.2.

9.3.1.2 Step 2: Winding Diagram of the PMSM

With ANSYS Motor-CAD, the PMSM winding diagram can also be readily implemented through the Winding menu. The lap winding is chosen from the menu. Its model can be easily completed by providing five winding parameters

TABLE 9.1

Stator Parameter

Stator Parameters	Values
Slot no	18
Armature diameter	80
Tooth width	3.5
Slot depth	9
Slot coroner radius	0
Slot opening	2
Tooth tip depth	1
Tooth tip angle	15
Air gap	1
Sleeve thickness	0
Axel diameter	58
Axel hole diameter	0

TABLE 9.2

Pole Parameter

Rotor Parameters	Values
Pole no	6
Back iron thickness	2
Magnet thickness	3.5
Magnet reduction	0.3918
Magnet arc	135
Magnet segment	1
Banding thickness	0

TABLE 9.3

Axial Parameter

Axial Dimension	Values
Motor length	120
Stator lam length	90
Magnet length	90
Magnet segments	1
Rotor lamination length	90
Ewdg overhang [F]	30
Ewdg overhang [R]	30
Wdg extension [F]	5
Wdg extension [R]	5
Shaft extension [F]	30
Shaft extension [R]	0

TABLE 9.4

Radial Parameter

Radial Dimension	Values
Stator lamination diameter	118
Stator bore	50
Banding thickness	0
Sleeve thickness	0
Magnetic thickness	0.35
Shaft diameter	25
Shaft diameter [F]	25
Shaft diameter [R]	25
Shaft hole diameter	25

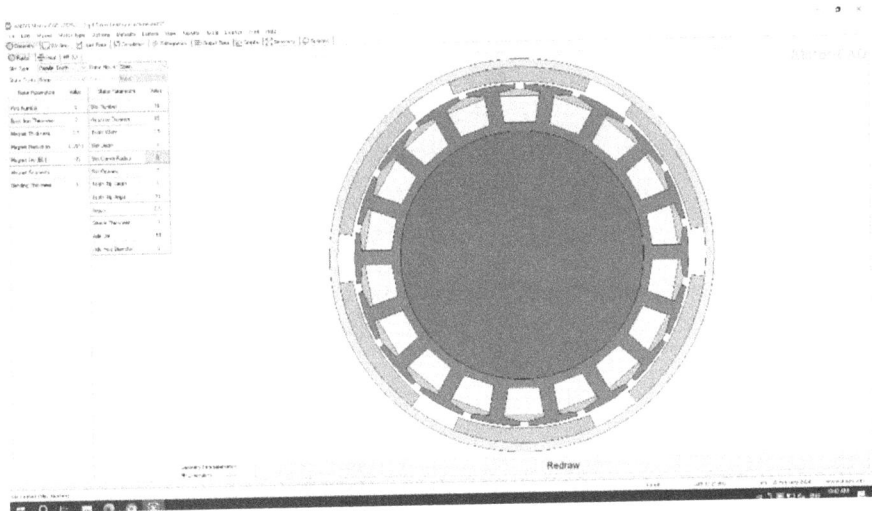

FIGURE 9.1
Outer rotor Surface Permanent Magnet (SPM) synchronous motor structure.

such as the number of phases, throw length in slot step, turns, and layers parallel pathways. The Pattern panel lets the user generate two types of windings automatically. These allow for an easy comparison between the modeled winding diagram and the one that the machine designer had previously created. The ANSYS Motor-CAD uses a radial arrangement of PMSM winding. The layered illustrations of the modeled (black lines) and designed iron developed PMSM winding for outer winding and inner winding are shown in Figures 9.3 and 9.4, respectively. Winding parameters of the PMSM are listed in Table 9.5.

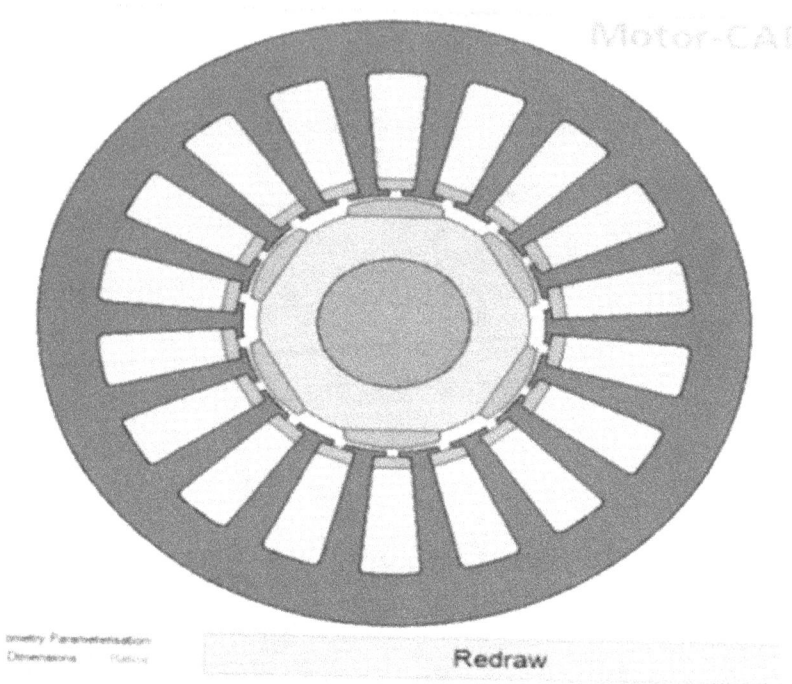

FIGURE 9.2
Inner rotor Surface Permanent Magnet (SPM) synchronous motor structure.

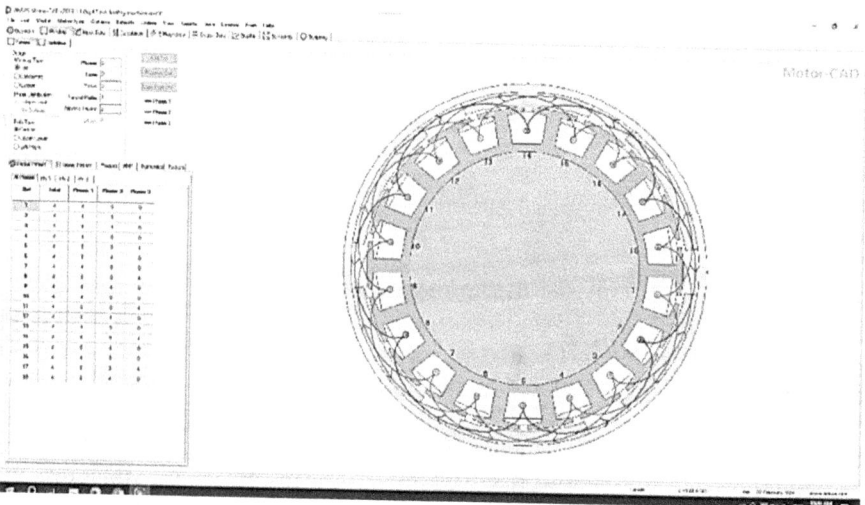

FIGURE 9.3
Outer rotor Surface Permanent Magnet (SPM) winding.

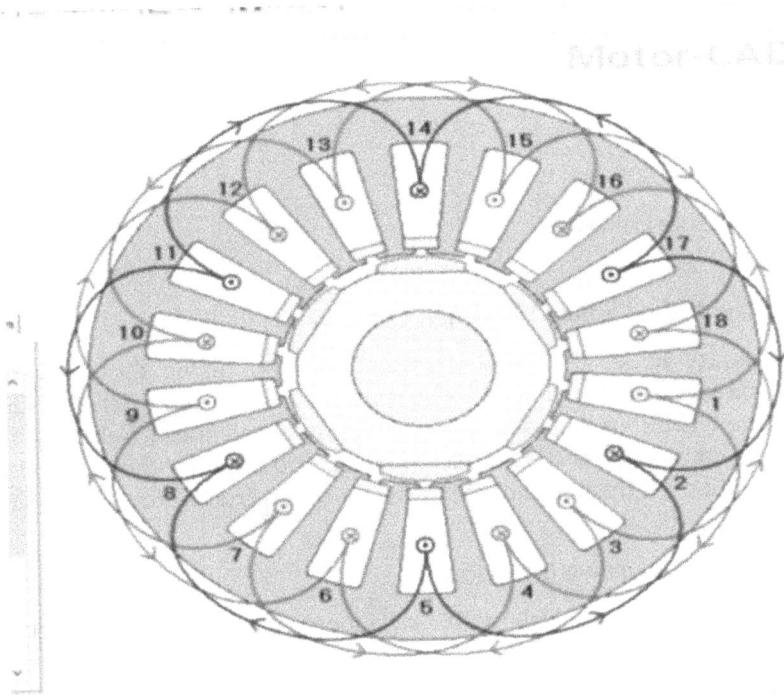

FIGURE 9.4
Inner rotor Surface Permanent Magnet (SPM) winding.

TABLE 9.5

Winding Parameters

Winding Parameters	Values
Winding layer	Double layer
Winding type	Lap
Parallel paths	1
Ewdg overhang [F]: End winding front of the motor	
Ewdg overhang [R]: End winding rear of the motor	
Wdg extension [F]: Front Winding extension	
Wdg extension [R]: Rare Winding extension	

9.3.1.3 Step 3: Assignment of Model Material Parameter

The materials needed to build the surface PMSM have to be determined in the subsequent modeling process. ANSYS Motor-CAD provides an extensive library of common materials used in electrical machinery for this purpose. Thankfully, the supplied database contained every material that was chosen for the PMSM's pre-sizing. The established database for the SPM is shown in Figure 9.5.

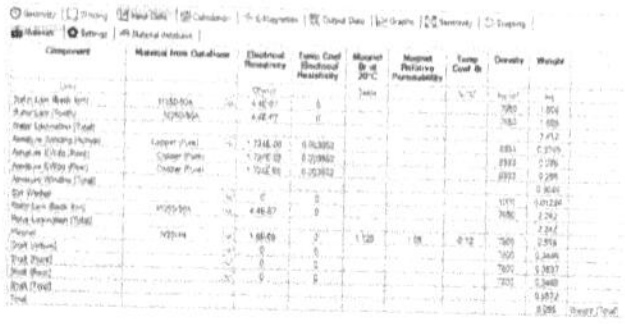

FIGURE 9.5
Materials needed for the model.

9.3.1.4 Step 4: Establishing Simulation Conditions

It is necessary to set up the simulation conditions using the software's Calculation menu. This panel includes configuration-related drive statistics, such as the DC bus voltage and the machine-imposed speed. Selecting the drive type, connecting the windings of stator, and aligning the magnetization orientations of the permanent magnets are also crucial steps.

9.3.1.5 Step 5: Simulating the Model

After completing the final phase, the development of the healthy machine model in ANSYS Motor-CAD was essentially complete. By selecting "solve electromagnetic model" and entering the appropriate simulation settings, the model is simulated.

9.3.1.6 Step 6: Creation of Air Gap Eccentricity

In Motor-CAD, the eccentricity can be modeled for all possible cases in the x–y plane. It is assumed that the stator and rotor axes in the z-direction are parallel, i.e., the bearings are not aligned.

The following eccentricity types are available:

- **Static Eccentricity**: Shifted stator axis, fixed rotor axis. The rotor axis center is at (0,0).
- **Dynamic Eccentricity**: Fixed stator axis, moving rotor axis. The stator axis center is at (0,0).
- **Static + Dynamic (Mixed) Eccentricity**: Shifted stator axis, moving rotor axis. Both canters are offset from (0,0).

The developed model is simulated under:

i. Healthy conditions
ii. Different static, dynamic, and mixed eccentricity circumstances at different levels. The findings are shown in the section that follows.

9.4 Results and Discussion

The complete machine structure can be simulated using the ANSYS Motor-CAD program, or just a part of it, due to the symmetry conditions in the model. The mesh fineness and calculation points are based on values of interest and other aspects, even if the FEA operates flawlessly with the solver's default settings. Numerous different outcomes were obtained from the FEA that was conducted. Only those that are most relevant to the study being conducted are described here. The completed FEA of the electrical machine model yields the fundamental results, which are displayed on the software's E-Magnetics panel. The current, voltage, back EMF, and torque waveforms of the healthy outer rotor Surface Permanent Magnet (SPM) synchronous motor and inner rotor Surface Permanent Magnet (SPM) synchronous motor are shown in Figures 9.6–9.9, respectively.

Table 9.6 presents the magnitude of torque harmonics for a healthy ORPMSM machine, whereas Tables 9.7–9.9 presents the values for the three eccentricities (Static, Dynamic, and Mixed) at levels of 10%–40% in steps of 10%. In Tables 9.6–9.9, 0th–15th refers to the order of harmonics, and 10%–40% refers to the level of harmonics.

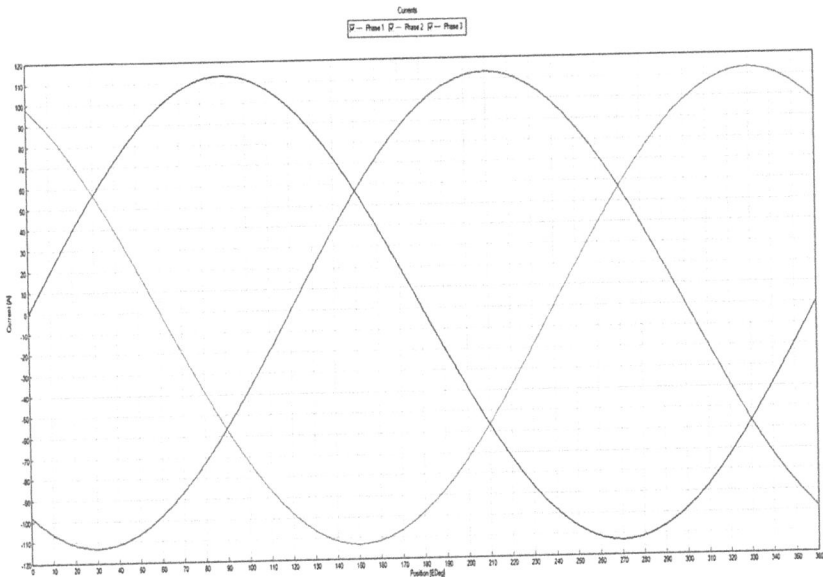

FIGURE 9.6
Current waveform.

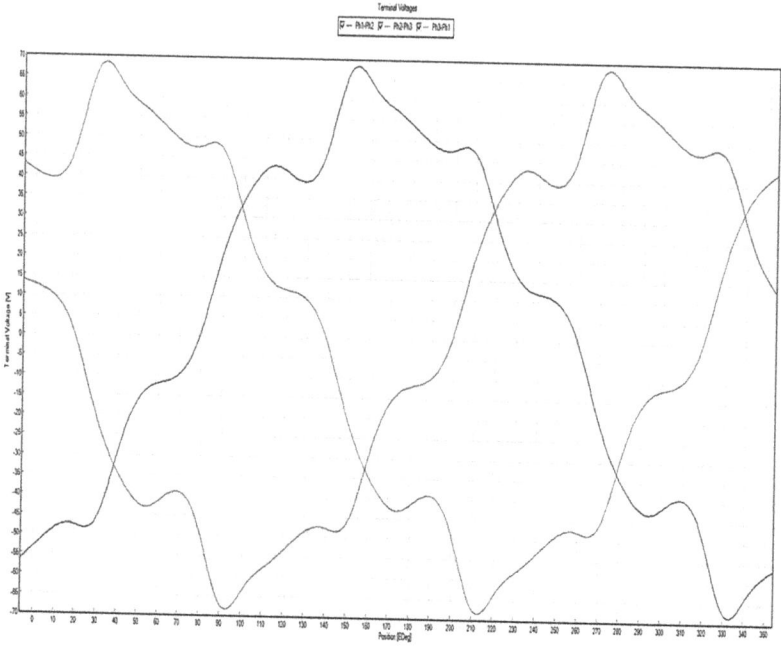

FIGURE 9.7
Terminal voltage (for an input DC voltage of 72 V).

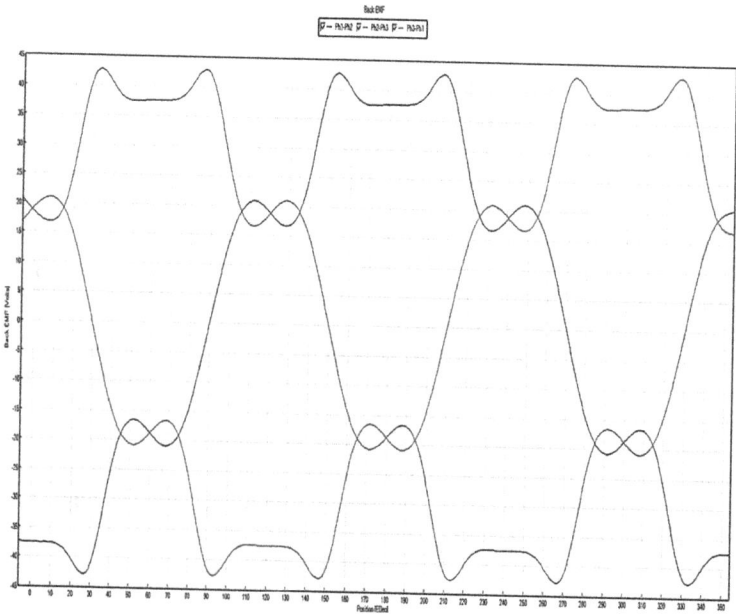

FIGURE 9.8
Back EMF.

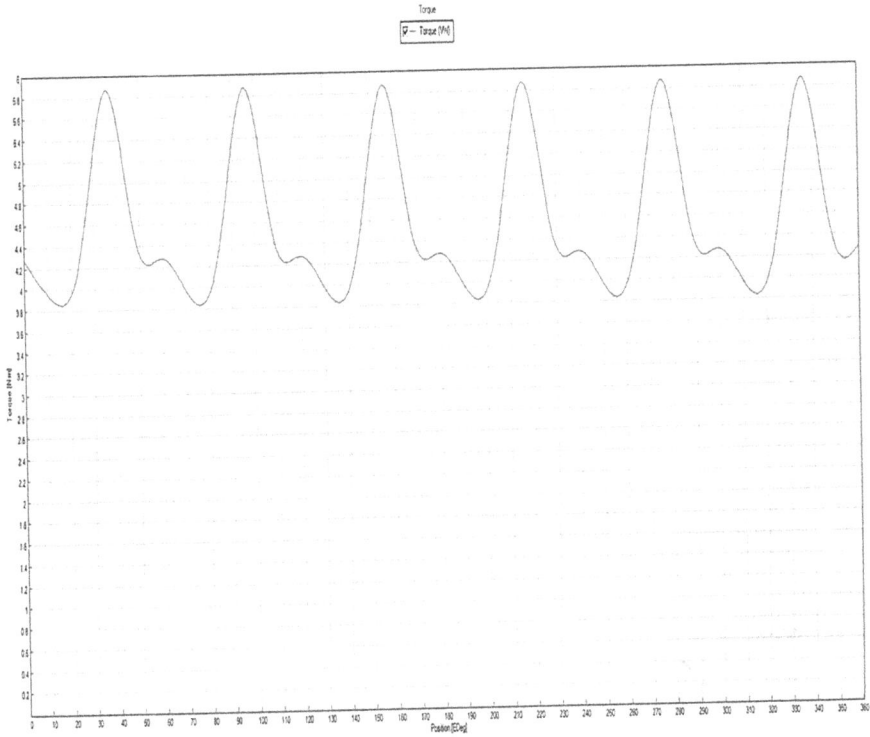

FIGURE 9.9
Torque (average torque=4.5 Nm).

TABLE 9.6

Healthy PMSM

0	1	2	3	4	5	6	7	8	9	10	11	12	13	14	15
4.568	0	0	0	0	0	0.795	0	0	0	0	0	0.521	0	0	0

TABLE 9.7

Static Eccentricity Harmonics

	0	1	2	3	4	5	6	7	8	9	10	11	12	13	14	15
10%	4.536	0	0	0	0	0	0.79	0	0	0	0	0	0.528	0	0	0
20%	3.756	0	0	0	0	0	1.642	0	0	0	0	0	0.431	0	0	0
30%	3.75	0	0	0	0	0	1.667	0	0	0	0	0	0.401	0	0	0
40%	3.41	0	0	0	0	0	1.672	0	0	0	0	0	0.391	0	0	0

TABLE 9.8

Dynamic Eccentricity Harmonics

	0	1	2	3	4	5	6	7	8	9	10	11	12	13	14	15
10%	4.536	0	0	0	0	0	0.79	0	0	0	0	0	0.528	0	0	0
20%	3.756	0	0	0	0	0	1.642	0	0	0	0	0	0.431	0	0	0
30%	3.75	0	0	0	0	0	1.667	0	0	0	0	0	0.401	0	0	0
40%	3.41	0	0	0	0	0	1.672	0	0	0	0	0	0.391	0	0	0

TABLE 9.9

Mixed Eccentricity Harmonics

	0	1	2	3	4	5	6	7	8	9	10	11	12	13	14	15
10%	3.389	0	0	0	0	0	0.911	0	0	0	0	0	0.621	0	0	0
20%	3.383	0	0	0	0.001	0	0.953	0	0	0	0	0	0.576	0	0	0
30%	3.184	0	0.001	0	0.002	0	1.084	0	0.001	0	0	0	0.543	0	0	0
40%	3.182	0	0.004	0	0.004	0	1.088	0	0.004	0	0.001	0.001	0.475	0	0	0

9.5 Conclusions

Motors used in electric vehicles are subjected to vibrations. As a result, bearing wear and tear is bound to occur. Hence, the rotor shaft position in the air gap will no longer be symmetrical. They may suffer from static eccentricity, dynamic eccentricity, and mixed eccentricity. From the model and simulated waveforms, it is observed that machine parameters such as torque, speed, and current are affected. From the detailed harmonic analysis of the torque waveforms for different eccentricity conditions, it can be inferred that "6n" harmonics (where n is the order of harmonics), along with the fundamental, are affected. Other harmonics are also affected, but the severity is less and hence can be ignored. It follows that the degree of air gap eccentricity in the machine can be determined non-invasively using harmonic analysis of the torque.

Acknowledgements

I convey my sincere gratitude to my guide Dr. B. L. Rajalakshmi Samaga, Professor, Department of Electrical and Electronics, NMAMIT, Nitte, Nitte University, and to the Centre for System Design (CSD) and Dr. Gangadhara at the National Institute of Technology Karnataka (NITK), Surathkal, for the technical assistance provided.

References

1. Samaga, BL Rajalakshmi, and Panduranga Vittal K. "Air gap mixed eccentricity severity detection in an induction motor." *2011 IEEE Recent Advances in Intelligent Computational Systems*. IEEE, 2011, Kerala, India.
2. Samaga, BL Rajalakshmi, and Panduranga Vittal K. "Comprehensive study of mixed eccentricity fault diagnosis in induction motors using signature analysis." *International Journal of Electrical Power & Energy Systems* 35.1 (2012): 180–185.
3. Samaga, BL Rajalakshmi, and Panduranga Vittal K. "A simplified modeling approach for accounting skewing effect in rotor bars of squirrel cage induction motor and its application in motor inductance calculation." *Journal of Electrical Engineering* 10.4 (2010): 7–7.
4. Coteț, Florina-Ambrozia, Iulia Văscan, and Loránd Szabó. "On the usefulness of employing ANSYS Motor-CAD software in designing permanent magnet synchronous machines." *Designs* 7.1 (2023): 7.
5. Bdewi, Mustafa Yaseen, Mohammed Moanes Ezzaldean Ali, and Ahmed Mahmood Mohammed. "In-wheel, outer rotor, permanent magnet synchronous motor design with improved torque density for electric vehicle applications." *International Journal of Electrical and Computer Engineering* 12.5 (2022): 4820.
6. Ayhan, Doğukan. "Analysis and optimization of interior permanent magnet synchronous motor for electric vehicle applications using ANSYS Motor-CAD." *International Journal of Automotive Engineering and Technologies* 12.3 (2023): 105–120.
7. Viswanath, Sreedharala, Praveen Kumar Nagarajan, and T. B. Isha. "Static eccentricity fault in induction motor drive using finite element method." *Advances in Electrical and Computer Technologies: Select Proceedings of ICAECT 2019*. Springer Singapore, Singapore, 2020.
8. Gherabi, Zakaria, et al. "Eccentricity fault diagnosis in PMSM using motor current signature analysis." *2019 International Aegean Conference on Electrical Machines and Power Electronics (ACEMP) & 2019 International Conference on Optimization of Electrical and Electronic Equipment (OPTIM)*. IEEE, Istanbul, Turkey, 2019.
9. Krichen, Manel, et al. "Effects of airgap static eccentricity in permanent magnet synchronous motors by means of finite element analysis." *2020 17th International Multi-Conference on Systems, Signals & Devices (SSD)*. IEEE, Sfax, Tunisia, 2020.
10. Rosero, Javier, et al. "Fault detection of eccentricity and bearing damage in a PMSM by means of wavelet transforms decomposition of the stator current." *2008 Twenty-Third Annual IEEE Applied Power Electronics Conference and Exposition*. IEEE, Austin, Texas, USA, 2008.
11. Wang, Lingling, et al. "Finite element analysis of permanent magnet synchronous motor of electric vehicle." *2015 3rd International Conference on Advances in Energy and Environmental Science*. Atlantis Press, Zhuhai, China, 2015.
12. Mersha, Tewodros Kassa, and Changqing Du. "Co-simulation and modeling of PMSM based on ANSYS software and Simulink for EVs." *World Electric Vehicle Journal* 13.1 (2021): 4.

13. Tikadar, Amitav, et al. "Coupled electro-thermal analysis of permanent magnet synchronous motor for electric vehicles." *2020 19th IEEE Intersociety Conference on Thermal and Thermomechanical Phenomena in Electronic Systems (ITherm).* IEEE, Florida, USAv 2020.
14. AKYÜN, Yakup, et al. "Design analysis and verification of PMSM motor for dishwasher machine." *2019 4th International Conference on Power Electronics and their Applications (ICPEA).* IEEE, Elazig, Turkey, 2019.

10

Deep Convolutional Neural Network-Based Smart Diagnostic System for Enhanced Breast Cancer Detection

Rajwinder Singh, Hardeep Kaur, Jyoteesh Malhotra,
Gajendra Kumar, and Komalpreet Kaur

10.1 Introduction

The occurrence of cancer cells in humans is very dangerous, as to date no permanent antidote has been developed for it. In the case of women, the growth of breast cancer is a serious issue and the second most common cause of mortality among them after lung cancer (Nassif et al., 2022). As documented in the 2023 report, WHO (World Health Organization) forecasted that between 2020 and 2040 there will be 2.5 million breast tumor deaths worldwide. By 2030, 25% of deaths due to breast cancer may be prevented globally if the mortality rate is reduced by 2.5% per year (WHO, 2023). There are different causes and risk factors associated with breast cancer (BC). Family history, late childbearing, alcohol intake, etc., are common reasons behind the onset of BC among females. The different phases of the emergence of BC are depicted in Table 10.1. However, with timely intervention and identification, many women worldwide can be saved from this deadly disorder. To cater to these requirements, artificial intelligence and machine learning have come into the picture. They play an important role in the early prediction and detection of such patients, thereby enhancing the survival chances of cancer-affected patients (Hady et al, 2020; Tang et al., 2021).

These days, deep learning (DL), which is based on the concept of ANN (artificial neural networks), is also being utilized for big datasets. DL architectures like RNN (recurrent neural network), DBN (deep belief network), CNN (convolutional neural network), and DNN (deep neural network) are commonly used in several domains such as image analysis, computer vision, speech synthesis, voice recognition, social media filtering, natural language processing, and bioengineering. Early identification of breast cancer can be achieved through the accurate prediction of the various phases of breast

DOI: 10.1201/9781003506478-10

TABLE 10.1

Breast Cancer Development Phases (Ara et al, 2021)

Phases	Overview	Dimension	Treatment Options
1	Non-invasive cancer	Tiny size	Simple mastectomy & BCS
2	Invasive	Less than or equal to 2 cm	Chemotherapy, hormone therapy, immunotherapy
3	Invasive	2–5 cm	Chemotherapy, hormone therapy, immunotherapy, and target drug
4	Invasive	>5 cm	Surgery, radiation therapy
5	Invasive	Any size	Hormone treatment, immunotherapy, and chemotherapy

cancer cells by ML and DL models, based on cancer size (Yash et al., 2022; Kononenko et al., 2001; Bataineh et al., 2019).

In this research work, three different ML models, namely RF (random forest), LR (logistic regression), and SVM (support vector machine), and two deep learning algorithms, viz. ANN and CNN, employed the WBCD (Wisconsin Breast Cancer Database) for testing and training to categorize breast cancer cases into either cancer or non-cancer groups. Various performance metrics are evaluated and compared to find the best-fit model for the accurate identification of breast carcinoma.

There are various segments in this study. Segment 1 describes BC and the associated causes and risks. Segment 2 lists the surveys relevant to this study. Segment 3 presents the description of the dataset used for the research work and the associated methodology, consisting of the depiction of preprocessing techniques, feature extraction methods, and a brief introduction to the applied ML and DL models. Segment 4 displays the research results and is followed by a work-related discussion. Segment 5 concludes the paper with the further scope of research.

10.2 Literature Survey

Different machine learning algorithms have been explored for the detection and diagnosis of BC by many authors to date using the SEER dataset, Wisconsin dataset, mammogram images, and real-time datasets from several hospitals. Relevant features were extracted and selected from these datasets, which were further fed as input to different ML models for detection purposes. In the past, many authors reviewed and summarized different traditional machine learning models that were being explored for breast cancer detection (Fatima et al., 2020; Keles 2019; Saba 2020). The authors (Delen et al., 2005) predicted breast cancer by comparing three data mining algorithms and demonstrated that the most accurate predictor was the DT, which had a success rate of 93.6%, followed by ANN and logistic regression with 91.2% and 89.2%, respectively.

In Sarvestani et al. (2010), the authors tried to solve the problem of the high dimensionality of the dataset by incorporating the PCA technique to minimize the data's size. Kharya et al. (2014) investigated and proposed that the NB classifier showed high testing accuracy with low computational time to detect breast cancer. In Asri et al. (2016), the performance comparison among Support Vector Machine (SVM), Decision Tree (DT), Naive Bayes (NB), and k-nearest neighbors (KNN) machine learning models was carried out using the WBCD. Various performance metrics, including recall, accuracy, precision, and F1 score, were computed. According to the results, out of all the models, the Support Vector Machine achieved the best accuracy (97.13%) with the least amount of error.

The authors (Osman 2017) proposed a smart detection scheme for breast cancer by using a hybrid SVM and the two-step clustering method. This built and trained hybrid method achieved a testing accuracy of 99.10% using the WBC dataset. The authors compared KNN and SVM machine learning algorithms to distinguish benign and malignant breast cancer categories using 3D images of FNA tissues to select and extract important features of the nuclei of the cells and found that the support vector machine model yielded better results with an accuracy of 97.49%.

In Vaka et al. (2020), a new method known as the deep neural network (DNN) for detecting breast cancer was proposed and compared with traditional ML models. The proposed DNN model provided better results than the existing methods. In Huang et al., (2020), the effectiveness of the AdaBoost machine learning model as a strong classifier was validated by using a large ultrasonic image dataset of 1,062 breast tumor cases with 418 benign and 644 malignant instances, and its performance was compared with different traditional models. The experimental outcomes revealed that the presented method provided the best classification results. Using BN and RBF algorithms, Jabbar (2021) created a collective model to diagnose breast cancer and achieved an accuracy of 97.42%. Further, Gopal et al. compared RF, LR, and MLP models using ten-fold cross-validation and PCA to predict BC. The results exhibited that MLP outperformed RF and LR models by achieving 98% classification accuracy.

Most studies have delved into machine learning applications utilizing the Wisconsin dataset. However, notable gaps persist in the literature. Some algorithms have yet to be tested, and there is a lack of comprehensive cross-study analyses of findings. From the above literature, it has also been observed that the deep learning domain has not been extensively explored on this dataset, as researchers have only focused on machine learning models.

Hence, in this paper, we have used both deep learning and machine learning models. It has been found that the results of deep learning models showed better performance than ML models. Hence, the adoption of deep learning models emerges as a solution to diagnose breast cancer. In the current research work, five ML and DL models were trained and tested on the WBCD dataset, and the key contributions of the study are presented below:

1. A computer-aided system to automatically detect breast cancer using different ML and DL approaches is developed.
2. Different performance metrics, viz., accuracy, sensitivity, specificity, precision, AUC-ROC, and F1 score for the considered models are compared to find the best-fit model.
3. A customized CNN-based deep learning model has exhibited the best results with exceptional classification accuracy of 100% to distinguish benign and malignant classes of BC.
4. When compared to other state-of-the-art alternatives employing the WBCD dataset, the proposed model showed better performance outcomes.

10.3 Dataset Acquisition and Procedures

This segment presents the methods utilized to conduct the entire study along with a description of the data employed for the research.

10.3.1 Dataset Acquisition

In this paper, the WBCD is used, which is openly available in the UC Irvine Machine Learning Repository (UCI Machine, 2022). The dataset was first used by Bennett and Mangasarian (1992) to categorize cancerous and non-cancerous (benign and malignant) cases. From digital pictures of the breast mass aspirated with a fine needle, many attributes were assessed. As seen in Figure 10.1, this data collection includes 212 cases of malignant (M) and 357 cases of benign (B) tumors. The dataset uses 32 attributes, which are shown in Table 10.2. A few essential features used to identify breast carcinoma are: radius, texture, borders, region, consistency, stiffness, concave patches, concaveness, harmony, and temporal scale.

10.3.2 Methodology

To conduct this study, the unprocessed BC dataset was first pre-processed and standardized. Next, the important features were extracted using feature selection and extraction techniques to provide them as input to ML and DL models to identify the best-fit model for the detection of BC. An outline of the steps taken to complete the research is provided in Figure 10.2.

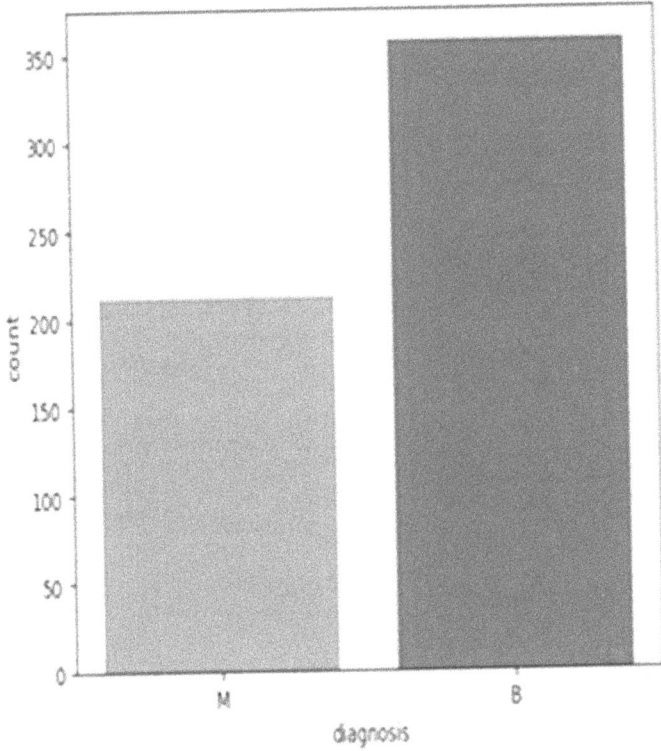

FIGURE 10.1
WBCD with the distribution of benign and malignant classes.

TABLE 10.2

Different Attributes of the WBCD (Kaur, 2023)

S.No.	Attribute	S.No.	Attribute	S.No.	Attribute
I	diagnosis	XII	radius_se	XXIII	texture_worst
II	Radius_mean	XIII	texture_se	XXIV	perimeter_worst
III	texture_mean	XIV	perimeter_se	XXV	area_worst
IV	perimeter_mean	XV	area_se	XXVI	smoothness_worst
V	Area_mean	XVI	Smoothness_se	XXVII	compactness_worst
VI	Smoothness_mean	XVII	compactness_se	XXVIII	concavity_worst
VII	Compactness_mean	XVIII	concavity_se	XXIX	Concave points_ worst
VIII	Concavity_mean	XIX	concave points_se	XXX	symmetry_worst
IX	Concave points_mean	XX	symmetry_se	XXXI	fractal_dimension_worst
X	Symmetry_mean	XXI	fractal_dimension_se	XXXII	id
XI	fractal_dimension_mean	XXII	radius_worst		

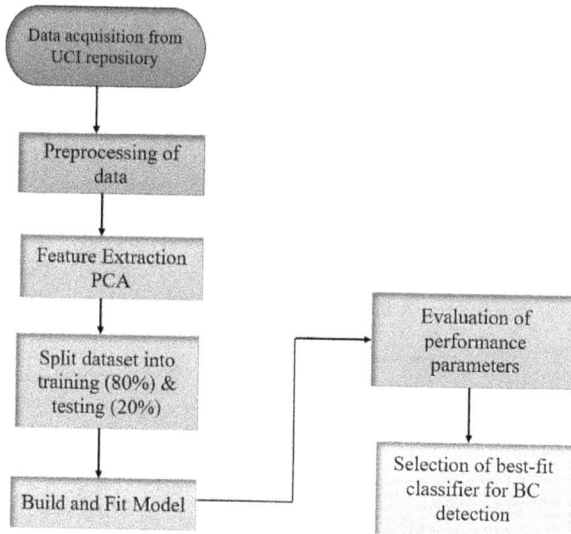

FIGURE 10.2
Flowchart of the methodology.

10.3.3 Preprocessing

The raw breast cancer dataset was preprocessed to remove different artifacts present in the dataset, followed by label encoding and standard scaling.
   For the preprocessing of the dataset, the following steps were executed:

- **Removal of Unwanted Aspects**: Unwanted aspects were first eliminated from the WBCD dataset during preprocessing. Since the feature "id" didn't offer any important information, it was eliminated from the collection.
- **Managing Erroneous Values**: Additionally, missing and null data values in the dataset were examined and processed by applying imputation methods using the mean, median, and mode.
- **Label Encoding**: Encoding was performed to convert the categorical data into numerical values for malignant and benign.
- **Scaling**: After label encoding, standardization of the attributes was carried out. A crucial preprocessing step, standardization is a subset of feature scaling. It requires rescaling the features to give them the characteristics of a Gaussian distribution with a zero mean and unit variance. Standardization is a basic requirement before training the machine learning models to achieve higher accuracy and maps the features to a standard Gaussian distribution based on $(h-r)/s$ where r and s are the mean and standard deviation, respectively, of the feature values.

- **Upsampling of Classes**: The cancerous and benign classes were prejudiced; hence, upsampling of the class with a lower count was carried out. This type of data augmentation involves balancing the counts of the two classes to prevent bias. Initially, the class distribution of the dataset was biased (212 malignant and 375 benign). After data augmentation, both the samples of cancer and non-cancer changed to the same size (375), as shown in Figure 10.3.

10.3.4 Feature Extraction

A statistical technique called PCA reduces the complexity of data with high dimensions. It achieves this by converting the initial variables into a new collection of variables known as the principal components, which are uncorrelated and responsible for most of the data's variability.

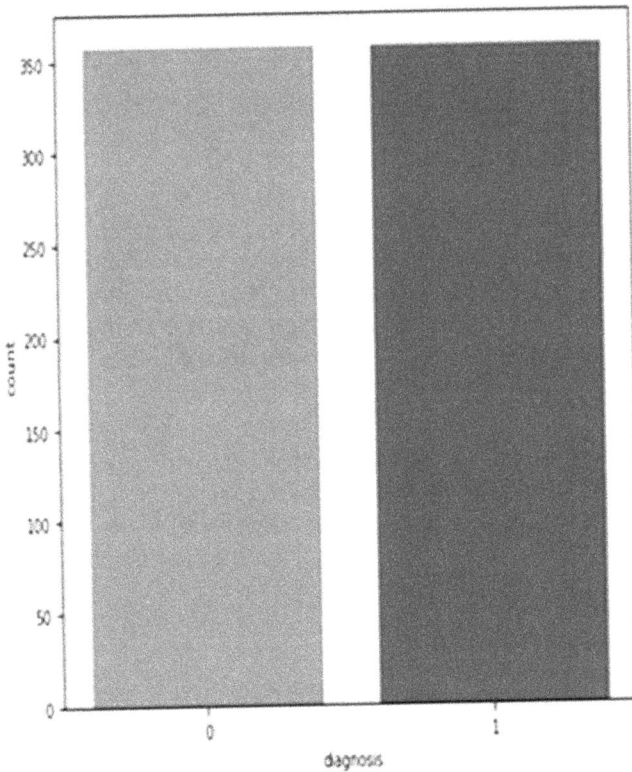

FIGURE 10.3
WBCD with benign and malignant cases distribution after data augmentation.

1. The first step in PCA is to create the covariance matrix, which calculates the correlations between every combination of variables included in the dataset.

2. The covariance matrix's independent values and self-vectors are ascertained in the second step. The directions of maximum variance are represented by the eigenvectors, and their magnitudes are shown by the eigenvalues.

3. After that, the eigenvalues are arranged in decreasing order in the third step. The associated eigenvector captures greater variation based on the higher eigenvalue.

4. The principal components (eigenvectors) that account for a substantial portion of the overall variance are selected in the fourth step.

5. Finally, to create a lower-dimensional representation, the initial data is superimposed into the chosen principal components.

Table 10.3 depicts the ten features extracted after PCA is executed.

10.3.5 Machine Learning Models

In this work, different ML models are used for the detection of benign and malignant breast tumor cells in patients. A brief description of these ML models is given below:

a. **Logistic Regression**: LR is a supervised machine learning algorithm specifically designed to address classification problems by modeling the probability of discrete outcomes. By categorizing, we imply that this model can be used to convert a group of input variables or features into desired values. In the scenario of logistic regression, the outcome is a probabilistic value between 0 and 1.

b. **Random Forest**: This model generates many interconnected decision chains. The backbone of this approach is decision trees. A collection of decision trees produced during the initial processing stage is referred to as a "random forest." After multiple trees are built, the best feature is randomly chosen from a list of attributes.

TABLE 10.3

Selected Features after PCA Execution

• radius_mean	• fractal_dimension_mean
• texture_mean	• smoothness_mean
• compactness_mean	• radius_se
• concave_points_mean	• texture_se
• symmetry_mean	• smoothness_se

c. **Support Vector Machine**: The Support Vector Machine (SVM) functions as a discriminative classifier, relying on the use of a splitting hyperplane. In n-dimensional space, a hyperplane denotes a flat subspace that may not intersect the origin and possesses $(n-1)$ dimensions. This concept of an $(n-1)$-dimensional flat subspace persists even when the hyperplane isn't explicitly visible in higher dimensions. When datasets lack a linearly separable hyperplane, constructing a linear classifier becomes impractical. To overcome this challenge, the kernel technique is applied, enabling the creation of nonlinear classifiers by using maximum-margin hyperplanes. Rather than depending solely on the dot product, a nonlinear kernel function is employed to define the hyperplanes.

d. **Artificial Neural Network**: The operational structure of neurons' dendrites, cell bodies, and axons lays the groundwork for artificial neural network (ANN) algorithms. Within an ANN, every component, from basic mathematical operations to artificial neurons, replicates the organization of real neurons. These networks consist of interconnected neurons grouped into input, hidden, and output layers, establishing the core architecture of ANNs. With accumulated experience, these networks become versatile tools for various tasks, particularly adept at tackling regression and classification challenges. Their widespread adoption in computer vision tasks highlights their specialization as deep learning algorithms in this field.

e. **Convolutional Neural Network**: Convolutional neural networks (CNNs) are based on the concept of backpropagation. They are quite useful for classifying and detecting objects. CNNs consist of multiple layers, such as fully connected, pooling, and convolutional layers. By constructing a framework comprising alternating layers of convolutional and pooling data, features are recovered and placed on a smooth layer [31]. As a result, CNNs can perform feature mapping automatically without the need for manual feature extraction techniques. The framework of CNNs consists of the following layers:

- **Convolution Layer**: The layer known as convolution is the foundation of any convolutional neural network model. It includes a group of filters called kernels fitted to the incoming data. The result of this layer is an attribute map. CNN layers extract features by employing linear and nonlinear processes. The mathematical representation of the convolution layer is (Thongsuwan & Jai, 2021):

$$(G \times H)_{x,y} = \sum_{p}\sum_{n} G_{x+p,y+l} \cdot H_{p,l} \qquad (10.1)$$

where G=input matrix; H, p, l=filter matrix; ×=convolution operation; x, y=output matrix

- **Pooling Stratum**: Pooling layers further minimize the quantity of initial data by employing two forms of pooling: maximum pooling and standard pooling. Pooling speeds up model training by reducing the number of resulting parameters. The pooling layer output is then passed through an entirely linked layer. Mathematical expression for pooling layer is (Thongsuwan & Jai, 2021):

$$Z_{xy} = \max_{ij} W_{x+i,y+j} \qquad (10.2)$$

where x, y=indices of output matrix; i, j=pooling region; Z=input feature map; W=pooled feature map

- **Completely Interlinked Layers**: The features learned by the convolution and pooling layers are accessed by the completely linked layer at the final stage of a CNN algorithm, enabling identification. The prior feature maps learned through the convolution and pooling layers are frequently flattened in the fully connected layer. The mathematical expression is given below (Thongsuwan & Jai, 2021)

$$z = K \cdot y + c \qquad (10.3)$$

whereas, z=output vector; K= weight matrix; y= input vector; c=bias vector

In the present study, Table 10.4 shows the hyper-tuning parameters employed for different models to execute the research work.

TABLE 10.4

Hyper-Tuning Parameters Used for Different Models

Models	Parameters
CNN	Activation function=relu, sigmoid, Kernal size=3, Epochs=200 Learning rate=0.001, Optimizer=Adam, Filters=16
ANN	Activation function=relu, sigmoid, Learning rate=0.001 Optimizer=Adam, Epochs=200
LR	Random state=51
RF	Estimators=20, Criterion=Entropy, Random state=51
SVM	Kernal=linear

10.3.6 Performance Metrics

After preprocessing, feature extraction, and application of ML models on the WBCD, confusion matrices were generated for all the considered models. The performance measurement values for each class, including accuracy, precision, sensitivity, and F1-score, are provided via a confusion array. A confusion matrix is a table with two dimensions, "Actual" and "Predicted," that indicates the relative positions of False Positive (FP), False Negative (FN), True Positive (TP), and True Negative (TN). These values generate several performance metrics for classification tasks (Sokolova et al., 2006; McHugh 2012; Pal et al., 2021). The performance parameters are mathematically given below:

$$\text{Accuracy} = \frac{TP + TN}{TP + FP + FN + TN} \tag{10.4}$$

$$\text{Precision} = \frac{TP}{TP + FP} \tag{10.5}$$

$$\text{Sensitivity} = \frac{TP}{TP + FN} \tag{10.6}$$

$$F1 - \text{Score} = 2 \times \frac{TP}{2} \times TP + FP + FN \tag{10.7}$$

$$\text{Cohens kappa } (K) = \frac{2 \times (TP \times TN - FN \times FP)}{(TP + FP) \times (FP + TN) + (TP + FN) \times (FN + TN)} \tag{10.8}$$

10.4 Results and Discussion

10.4.1 The Experimental Results of the Different Models for Breast Cancer Detection

Table 10.5 and Figure 10.4 demonstrate the performance of classical ML/DL classifiers in terms of accuracy, precision, F1 score, recall, Cohen's kappa, and AUC-ROC using tested samples for breast cancer detection. DL models were trained using 200 epochs. The CNN model showed exceptional outcomes by achieving an accuracy of 100%, followed by ANN, random forest, logistic regression, and SVM with 99.44%, 99%, 98.80%, and 98.30%, respectively. Four models, namely, CNN, ANN, random forest, and logistic regression, achieved a precision score of 1.00, while SVM exhibited a score of

TABLE 10.5

The Experimental Findings of the Different Models for Breast Cancer Detection

Performance Statistics	CNN	ANN	LR	RF	SVM
Accuracy	1.00	0.994	0.988	0.99	0.983
Precision	1.00	1.00	1.00	1.00	0.989
F1-Score	1.00	0.995	0.989	0.989	0.984
Recall	1.00	0.989	0.978	0.978	0.978
Cohens Kappa	1.00	0.989	0.978	0.978	0.966
AUC-ROC	1.00	0.995	0.989	0.989	0.986

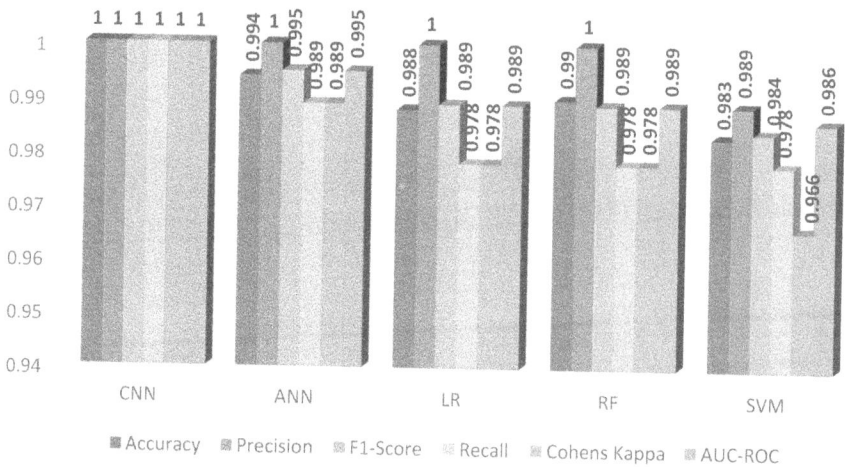

FIGURE 10.4
The performance metrics of the different models for breast cancer detection.

0.989. The CNN model also outperformed the other models by achieving an F1 score of 1.00, a recall score of 1.00, a Cohen's kappa score of 1.00, and an AUC-ROC of 1.00, which proves the efficacy of the CNN-based DL model for the detection of breast cancer.

10.4.2 Confusion Matrix for Proposed Model

The confusion matrix of the proposed CNN model is shown in Figure 10.5. The proposed model correctly predicted breast cancer for both classes. It shows that there are no false-negative and false-positive cases.

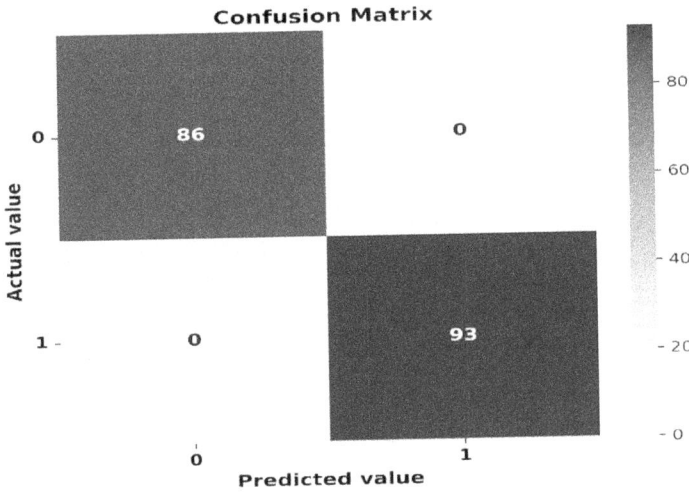

FIGURE 10.5
Confusion matrix of CNN model.

10.4.3 Accuracy and Loss Curve for Training and Validation of the Best Fit Proposed CNN Model

Figure 10.6 shows the accuracy curve and loss curve for training and validation subsets employing 200 epochs for the diagnosis of breast cancer. Figure 10.6a reveals that, for training and validation subsets, the accuracy curve increases exponentially with the number of epochs until it reaches nearly maximal accuracy. The loss curve in Figure 10.6b depicts an exponential reduction for both training and validation subsets as the number of epochs increases. It eventually converges to a minimum loss value. The accuracy and loss function curves show that the suggested CNN architecture is reliable and suitable for detecting breast cancer patients with great precision.

10.4.4 Comparison of the Proposed Model with Existing Research Work

A comparison of the suggested model is shown with earlier reported studies on the WBCD dataset for BC detection in Table 10.6. It is observed that the proposed model outperformed the other mentioned works by yielding 100% accuracy. Ahmed et al. attained an accuracy of 99.68% by utilizing a deep belief network (DBN-NN). The artificial meta plasticity multilayer perceptron (AMMLP) technique, with an accuracy of 99.26%, was proposed by Marcano-Cedeo et al. (2011) for the identification of breast cancer. Sonal Jain et al. attained 92% accuracy by using a k-means classifier. In (Aalaei et al., 2016;

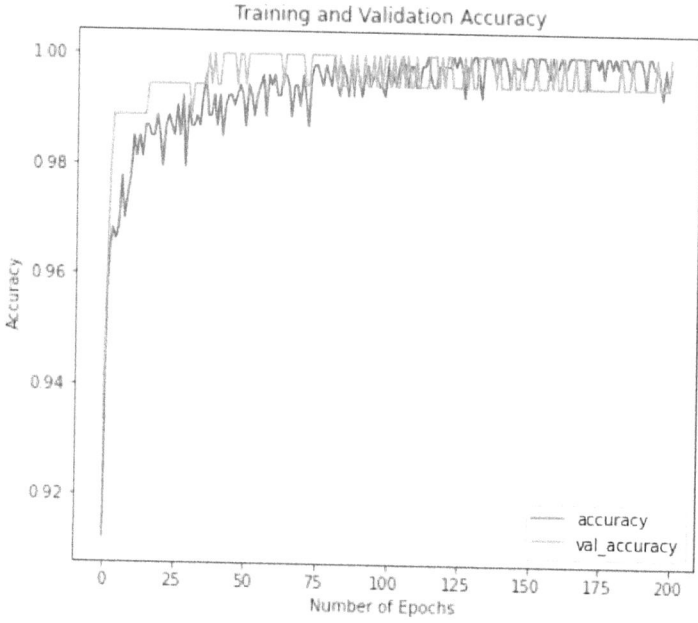

(a)

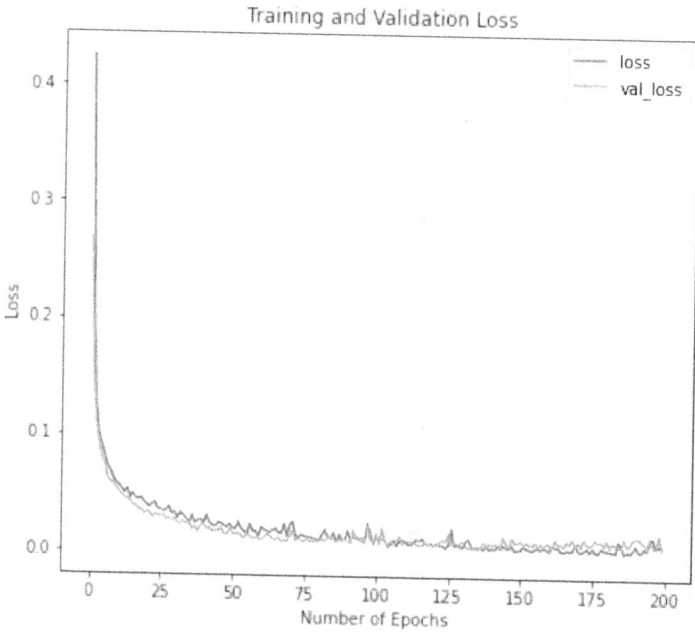

(b)

FIGURE 10.6

(a) Accuracy curve of CNN model, (b) loss curve of CNN model.

TABLE 10.6

Comparison of the Proposed Model with Existing Studies on WBCD

Authors	Year	Classifier	Accuracy
Marcano-Cedeno et al. (2016)	2016	AMMLP	99.2%
Saeid Eslami et al. (2016)	2016	ANN	97.3%
Asri et al. (2016)	2016	SVM	97.13%
Shawrarib et al. (2020)	2020	ANN	99.57%
Alshayeji et al. (2022)	2022	ANN	99.85%
Proposed model	2024	CNN	100%

Shawrarib et al., 2020), ANN classifier was used to achieve accuracy levels of 97.3%, 99.57%, and 99.85%, respectively. Asri et al. (2016) acquired 97.13% accuracy by using an SVM classifier to identify breast cancer cells. Therefore, the proposed CNN-based model shows its effectiveness in breast cancer detection by yielding 100% classification accuracy.

10.5 Conclusion

In this research work, different machine learning and deep learning classifier models such as LR, RF, SVM, ANN, and CNN were applied to the WBCD to identify benign and malignant cases in patients with breast cancer. Various performance measures, including the F1 score, AUC-ROC, recall, sensitivity, accuracy, precision, and Cohen's kappa, were assessed and contrasted with other ML/DL algorithms to identify the best-fit algorithm that can detect breast tumors among patients with high accuracy. It is concluded that the proposed CNN model outperformed others by achieving an accuracy of 100%, a precision of 1.00, an F1 score of 1.00, and an AUC-ROC of 1.00. Hence, CNN-based models can be used for the detection of BC among female patients with remarkable accuracy.

References

M. H. Alshayeji, H. Ellethy, S. Abed, and R. Gupta, (2022). Computer-aided detection of breast cancer on the Wisconsin dataset: An artificial neural networks approach, *Biomedical Signal Processing and Control*, Vol. 71, No. Part A, 103141. ISSN 17468094, https://doi.org/10.1016/j.bspc.2021.103141

S. Ara, A. Das, and A. Dey, (2021). Malignant and Benign breast cancer classification using machine learning algorithms, in *2021 International Conference on Artificial Intelligence (ICAI)*, no. June, 2021, pp. 97–101. https://doi.org/10.1109/ICAI52203.2021.9445249.

H. Asri, H. Mousannif, H.A. Moatassime, and T. Noel, (2016). Using machine learning algorithms for breast cancer risk prediction and diagnosis, *Procedia Computer Science*, Vol. 83, pp. 1064–1069. https://doi.org/10.1016/j.procs.2016.04.224

A. A. A Bataineh, (2019). Comparative analysis of nonlinear machine learning algorithms for breast cancer detection, *International Journal of Machine Learning and Computing*, Vol. 9, No. 3, pp. 248–254. https://doi.org/10.18178/ijmlc.2019.9.3.794

K. P. Bennett and O. L. Mangasarian, (1992). Robust linear programming discrimination of two linearly inseparable sets, *Optimization Methods and Software*, Vol. 1, No. 1, pp. 23–34. https://doi.org/10.1080/10556789208805504

D. Delen, G. Walker, and A. Kadam, (2005). Predicting breast cancer survivability: A comparison of three data mining methods, *Artificial Intelligence in Medicine*. Vol. 34, No. 2, pp. 113–127. https://doi.org/10.1016/j.artmed.2004.07.002

Fatima, L. Liu, S. Hong, and H. Ahmed, (2020). Prediction of breast cancer, comparative review of machine learning techniques, and their analysis, *IEEE Access*, Vol. 8, pp. 150360–150376. https://doi.org/10.1109/ACCESS.2020.3016715

V. N. Gopal, F. Al-Turjman, R. Kumar, L. Anand, and M. Rajesh, (2021). Feature selection and classification in breast cancer prediction using IoT and machine learning, *Measurement*, Vol. 178, 109442, https://doi.org/10.1016/j.measurement.2021.109442

A. Hady, A. Ghubaish, T. Salman, D. Unal, and R. Jain, (2020). Intrusion detection system for healthcare systems using medical and network data: A comparison study, *IEEE Access*, Vol. 8, pp. 106576–106584. https://doi.org/10.1109/ACCESS.2020.3000421

Q. Huang, Y. Chen, L. Liu, D. Tao, and X. Li, (2020). On combining biclustering mining and adaboost for breast tumor classification, *IEEE Transactions on Knowledge and Data Engineering*, Vol. 32, No. 4, pp. 728–738. https://doi.org/10.1109/TKDE.2019.2891622

M. A. Jabbar, (2021). Breast cancer data classification using ensemble machine learning, *Engineering and Applied Science Research*, Vol. 48, No. 1, pp. 65–72. https://doi.org/10.14456/easr.2021.8

H. Kaur, (2023). Dense convolutional neural network based deep learning framework for the diagnosis of breast cancer. *Wireless Personal Communications*. https://doi.org/10.1007/s11277-023-10678-9

M. K. Keles, (2019). Breast cancer prediction and detection using data mining classification algorithms: A comparative study, *Tehnički vjesnik*, Vol. 26, No. 1, pp. 149–155. https://doi.org/10.17559/TV-20180417102943

S. Kharya, S. Agrawal, and S. Soni, (2014). Naïve Bayes classifiers: Probabilistic detection model for breast cancer, *International Journal of Computer Applications*, Vol. 92, No. 10, pp. 26–31. https://doi.org/10.5120/16045-5206

I. Kononenko, (2001). Machine learning for medical diagnosis: History, state of the art and perspective, *Artificial Intelligence in Medicine*, Vol. 23, No. 1, pp. 89–109. https://doi.org/10.1016/S0933-3657(01)00077-X

A. Marcano-Cedeño, J. Quintanilla-Domínguez, and D. Andina, (2011). WBCD breast cancer database classification applying artificial metaplasticity neural network, *Expert Systems with Applications*, Vol. 38, No. 8, pp. 9573–9579. ISSN 0957-4174. https://doi.org/10.1016/j.eswa.2011.01.167

M. L. McHugh, (2012). Interrater reliability: The kappa statistic. *Biochemia Medica (Zagreb)*, Vol. 22, No. 3, pp. 276–82. PMID: 23092060; PMCID: PMC3900052.

A. B. Nassif, M. A. Talib, Q. Nasir, Y. Afadar, and O. Elgendy, (2022). Breast cancer detection using artificial intelligence techniques: A systematic literature review, *Artificial Intelligence in Medicine*, Vol. 127. https://doi.org/10.1016/j.artmed.2022.102276

A. H. Osman, (2017). An enhanced breast cancer diagnosis scheme based on two-step-SVM technique, *International Journal of Advanced Computer Science and Applications*, Vol. 8, No. 4, pp. 158–165. https://doi.org/10.14569/IJACSA.2017.080423

A. Pal, (2021). Logistic regression: A simple primer, *Cancer Research, Statistics, and Treatment*, Vol. 4, No. 3, pp. 551–554. https://doi.org/10.4103/crst.crst_164_21

T. Saba, (2020). Recent advancement in cancer detection using machine learning: Systematic survey of decades, comparisons and challenges, *Journal of Infection and Public Health*, Vol. 13, No. 9, pp. 1274–1289. https://doi.org/10.1016/j.jiph.2020.06.033

A. S. Sarvestani, A. A. Safavi, N. M. Parandeh, and M. Salehi, (2010). Predicting breast cancer survivability using data mining techniques, in *Proceedings of the 2nd International Conference on Software Technology and Engineering*, V2–227-V2–231. https://doi.org/10.1109/ICSTE.2010.5608818

M. Z. A. Shawrarib, A. E. B. Latif, B. E. E. Al-Zatmah, and S. S. Abu-Naser, (2020). Breast cancer diagnosis and survival prediction using JNN. *International Journal of Engineering and Information Systems (IJEAIS)*, Vol. 4, No. 10, pp. 23–30.

M. Sokolova, N. Japkowicz, and S. Szpakowicz, (2006). Beyond accuracy, F-score and ROC: A family of discriminant measures for performance evaluation, *AI 2006: Advances in Artificial Intelligence, Lecture Notes in Computer Science*, Vol. 4304, pp. 1015–1021. https://doi.org/10.1007/11941439_114

K. J. W. Tang, C. K. E. Ang, T. Constantinides, V. Rajinikanth, U.R. Acharya, K.H. Cheong, (2021). Artificial intelligence and machine learning in emergency medicine, *Biocybernetics and Biomedical Engineering*, Vol. 41, No. 1, pp. 156–172. https://doi.org/10.1016/j.bbe.2020.12.002

S. Thongsuwan and S. Jai, (2021), ConvXGB: A new deep learning model for classification problems based on CNN and XGBoost, *Nuclear Engineering and Technology*, Vol. 53, No. 2, pp. 522–531. ISSN 1738-5733. https://doi.org/10.1016/j.net.2020.04.008

UCI Machine Learning Repository: Breast Cancer Wisconsin (Diagnostic) Data Set. Accessed on 2 Jan (2022).

A. R. Vaka, B. Soni, and S. Reddy K., (2020). Breast cancer detection by leveraging Machine Learning, *ICT Express*, Vol. 6, No. 4, pp. 320–324. https://doi.org/10.1016/j.icte.2020.04.009

World Health Organization, WHO-Breast Cancer. Available at: https://www.who.int/news-room/fact-sheets/detail/breast-cancer. Accessed on 10 July (2023).

A. Yash, P. Pipariya, S. Patel, and M. Shah, (2022). Comparative analysis of breast cancer detection using machine learning and biosensors, *Intelligent Medicine*, Vol. 2, No. 2, pp. 69–81. https://doi.org/10.1016/j.imed.2021.08.004

Index